Natural Philosophy in the Graduation Theses of the Scottish Universities in the first half of the seventeenth century

D1618554

Giovanni Gellera

Contents

List of abbreviations

AT: R. Descartes, *Oeuvres*, C. Adam - P. Tannery (eds.), Paris 1897-1910;
CG: Thomas Aquinas, *Summa contra Gentiles*;
DM: F. Suárez, *Disputationes metaphysicae*, Cologne 1597;
DNB: *The Oxford Dictionary of National Biography*, Oxford University Press, 2004-2012, http://www.oxforddnb.com;
FAM: P. J. Anderson (ed.), *Fasti Academiae Mariscallanae Aberdonensis*, Aberdeen, The New Spalding Club, vol. II, 1898;
In octo Physic.: Thomas Aquinas, *Commentaria in octo libros Physicorum*;
In Phys.: College of Coimbra, *In octo libros physicorum Aristotelis Stagiritae partes duae*, Cologne 1625;
Met.: Aristotle, *Metafisica*, G. Reale (ed.), Milano 1995;
OG: P. J. Anderson, *Officers and Graduates of University and King's College, Aberdeen, MVD-MDCCCLX*, Aberdeen, The New Spalding Club, 1893;
Phys.: Aristotle, *Physics*;
SPhQ: Eustachius a Sancto Paulo, *Summa Philosophiae quadripartita, de rebus dialecticis, ethicis, physicis, & metaphysicis*, Cambridge 1640;
ST: Thomas Aquinas, *Summa theologiae*;
TL: *Theses logicae*;
TM: *Theses metaphysicae*;
TP: *Theses physicae*.

Introduction

I shall investigate the natural philosophy of the graduation theses of the Scottish universities in the first half of the seventeenth century. I shall seek to prove that the natural philosophy of the Scottish universities can be defined as 'Eclectic Scotistic Reformed Scholasticism'. The focus will be on two concepts of general physics: prime matter and movement. These concepts are fundamental to the understanding of Scholastic natural philosophy and its relation to early modern philosophy and science. My primary focus will be on the former aspect.

1. Natural philosophy in the graduation theses in Scotland in the first half of the seventeenth century

In the first half of the seventeenth century the academic teaching in Scotland was still conducted according to the Scholastic way, inherited from the Medieval Scholasticism of the twelfth and thirteenth centuries. This movement was still strong across Europe in the early modern age, and the Scottish universities are a part of this narrative. Scholasticism is the historical product of the attempted harmonisation of two great philosophical traditions with the Christian revelation: on the one side, Aristotelianism, the long-established tradition of commentaries and interpretations of the corpus of Aristotle, which flourished again in the thirteenth century in virtue of the European reception of the Arabic commentary tradition. On the other hand, Augustinianism, the philosophy inspired by Saint Augustine, more closely related to the Platonic tradition.

From the late Middle Ages to the early modern era, Scholasticism underwent deep changes: as has been argued by Charles Schmitt regarding Aristotelianism, it is more accurate to talk of 'Scholasticism*s*' rather than 'Scholasticism' as a monolithic body. Scholasticism is divisible into different schools (Thomism, Scotism and Nominalism, just to name the most important ones) and into different disciplines (Scholastic philosophy and

Scholastic theology). Scholasticism was also influenced by Renaissance philosophy, Renaissance Aristotelianism and Humanism, and was finally challenged by the rise of the new science in the early seventeenth century. A key aspect of my research is the investigation of the Scottish graduation theses in relation to the history of Scholasticism. I shall argue that the philosophy of the graduation theses is Scholastic in nature, heavily influenced by Scotistic themes, yet enriched by an eclectic character.

It cannot be forgotten that Scholasticism was born as an enterprise of human (or "natural") reason to penetrate the mysteries of the revelation: *'intellectus quaerens fidem'*, intelligence in search of faith, but also *'fides quaerens intellectum'*, faith in search of intelligence, according to the famous phrase of Anselm of Canterbury. Thus, the history of Scholasticism is also, at least partially, the history of the European Christian faith up to the modern age. The historical evolution of the Roman Church first, and later on of the Reformed churches played a major role in the development of Scholasticism, in terms of different schools, traditions and doctrines. I shall seek to investigate graduation theses from this point of view as well, in order to assess whether the Scottish Reformation influenced the Scholastic philosophy taught in Scotland. My answer will be that Scottish Scholasticism can be properly qualified as 'Reformed Scholasticism'.

Natural philosophy is the discipline which investigates natural bodies: their principles, properties and structure. Following Aristotle, 'natural bodies' are defined as bodies endowed with a 'nature', understood as the internal principle of change or movement in general. Natural bodies are thus defined by their nature, and change, that is any passage from potency to act, is the first and main consequence of their nature.

Natural philosophy is divided into general physics and special physics: the former deals with the general principles of bodies, namely, what qualifies them in general qua bodies. The latter is a cluster of different disciplines (for example, astronomy, chemistry, psychology), and deals with bodies as they fall into these disciplines. General physics extends further than any discipline of special physics, and includes them all as the genus includes the species.

My focus will be on two theories of general physics: prime matter and movement. Prime matter is the material principle of all bodies. In the framework of the Aristotelian theory of cause, prime matter is the material cause of bodies, a constitutive principle of the natural body, when united with the formal cause (form). Prime matter is traditionally and famously defined as 'pure potency'. We will see how the regents reject the Thomistic theory, and side with Duns Scotus.

The Scholastic theory of movement is central to our understanding of the very nature of Scholastic natural philosophy, precisely because of the close relation between nature and movement. Nature is the principle of movement, therefore movement follows from the nature of bodies, and informs us about it. We will see that the regents are still committed to Scholastic natural philosophy, yet in an original sense.

Prime matter and movement are the two key concepts of Scholastic natural philosophy. The analysis of them helps to clarify general physics as a whole, since prime matter and movement are not intelligible without the broader context of theories such as act and potency, form and matter, substance and the four Aristotelian causes. Regarding prime matter, I shall argue that graduation theses reveal the influence of Scotistic themes, as they deploy a metaphysics of essence and the notion of metaphysical (or entitative) act. Regarding movement, graduation theses inscribe themselves in the Scholastic tradition of natural places, directedness of movement and difference in nature between sublunary and celestial bodies.

The choice of prime matter and movement is also motivated by historical considerations. The seventeenth century saw the rise of the new science and the consequent revolution in our understanding of the world. Bacon, Copernicus, Galileo and Descartes are just the main figures of this revolution in the first half of the seventeenth century. As a consequence, the Scholastic notions of prime matter and movement were extensively discussed and criticised in the first half of the seventeenth century. Although they are still within the Scholastic tradition, graduation theses bear witness to this debate: we will see that some theses break with Scholasticism and, more generally, that the form of Scholasticism in use in the Scottish universities seems to anticipate later themes in early modern philosophy. I shall seek to highlight these aspects, especially in the theory of natural places and secondary causes, the definition of accident and the theory of substance.

Yet, my approach aims at shedding light not on the graduation theses in relation to the so-called 'modern philosophy' or 'early modern philosophy', but instead on the graduation theses within the Scholastic tradition; and I shall account for their specific character in the light of this tradition. Nonetheless, I believe that graduation theses not only anticipate some themes of modern philosophy but also that, more generally, Scholasticism in the seventeenth century prepared the ground for modern philosophy in a way that is yet to be fully acknowledged by scholars. While keeping the focus on Scottish Scholasticism, I hope to shed some light on these connections.

An important consequence of my approach is the choice of the period that I shall investigate. Regarding the terminus a quo, the first printed set of graduation theses available, which is by J. Robertson, regent at the University of Edinburgh, is dated 1596. Before that date, there is no printed evidence of what exactly was taught in the universities. The terminus ad quem requires more justification. I shall examine graduation theses until

1649, the latest being that of D. Forrester 1649, again from the University of Edinburgh. Up to that date, university teaching was still fully Scholastic in form and contents, while from the 1650s onwards we witness the epoch-making beginning of the reception of Descartes' philosophy in Scotland: Andrew Cant's *Theses philosophicae*, Aberdeen 1654, written for Marischal College, are the first graduation theses which refer to Descartes.

The reception of Descartes and, thereafter, of other modern philosophers, produced a profound change in the philosophy of the Scottish universities, which ultimately led to the shift from Scholasticism to the Enlightenment and modern science, less than one century later. I shall argue that the 1640s are the final years of the long tradition of Scottish Scholasticism in its purer form. This does not mean that this tradition did not survive the arrival of modern philosophy: as a matter of fact, Scholasticism was influential for the whole century and ultimately shaped the reception of modern philosophy in Scotland. I shall hint at this historical fact by reference to, for example, the theory of the relation between prime matter and quantity: the Cartesian notion of *res extensa* was quickly received in the Scottish universities in the 1660s and 1670s because it was anticipated by the Scottish Scholastic concept of quantity, one of the properties of prime matter, as essentially extended in place.[1] A similar point could be made regarding separate substances as the object of metaphysics, the concept of mind and of its faculties, the role of perception, just to name some other philosophical theories common in Scottish Scholasticism.[2]

One final premise of my research is the idea that the corpus of graduation theses of the first half of the seventeenth century can be investigated as a uniform, collective philosophy. From the point of view of the historical unity of my sources, graduation theses

---

[1] I believe that in the graduation theses the expression 'extension in place' is equivalent to 'extension in space'. I prefer the former because it is closer to the original Latin, *extensio in ordine ad locum*. Every place (*locus*) is spatially extended, while not every space is a place (for example, the *spatium imaginarium* around the upper limit of the heavens). The regents favour 'place' instead of 'space', while Suárez some times favours 'space', as in *DM*, 40, II, 22, '*extensio in spatio*'. As it appears in part I, chapter 4, section 4 regarding the distinction between *extensio in ordine ad se* and *extensio in ordine ad locum*, one of the characteristic of a place is that it is extended in space. Baron 1627, TP IV, explains the relation between place and space: *"Forma, quae motu locali acquiratur, et ejus terminus ad quem est* Ubi, *realis praesentia rei in loco, sive vero, sive imaginario spatio."* Therefore, even if the expression 'extension in space' is perhaps more intuitive than 'extension in place', I follow more closely the original text.

[2] *"The thesis, then, is not that the seventeenth-century brand of Scholasticism directly influenced Descartes' formulation of his philosophy but that, at least, it prepared the way for the acceptance of Cartesianism."* R. Ariew - M. Grene, in *Descartes and the Last Scholastics*, Ithaca - London, Cornell University Press, 1999, p. 78, fn. 1. Ariew and Grene admittedly expand a thesis by V. Carraud of the presence of a form of Ockhamism in Arnauld's philosophy, which drew him close to Descartes' *Meditationes*. V. Carraud, *Arnauld: From Ockhamism to Cartesianism*, in R. Ariew - M. Grene (eds.), *Descartes and his Contemporaries*, Chicago - London, University of Chicago Press, 1995, pp. 110-128. Regarding the Scholasticism of the graduation theses, I agree with the general view that some specific forms of Scholasticism were perceived to be close to Cartesian doctrines.

are the same type of text across all of Scotland, arguably written in similar material, cultural and social conditions. It is thus possible to investigate them under the collective name of 'graduation theses' of the Scottish universities. But the claim that their philosophy can be treated collectively, or *sub specie scholae*, is more debatable. I shall seek to investigate graduation theses as a philosophical unity, not simply as a historical one: that is, I shall seek to prove that they introduce us to a common philosophy characterised by the acceptance of an identifiable set of key doctrines. It is not enough to describe graduation theses as 'Scholastic', and the more precise term 'Scotistic' falls short of the target as well. Therefore, graduation theses are a unitary body whether regarded on their own or in relation to the Scholastic tradition. The risk of my approach is that the individual contribution may be overlooked in favour of the general notion of the 'philosophy of the graduation theses'; the opposite risk is to underestimate the general acceptance of some theories in the name of the respect for the individual philosopher. I shall seek to balance this collective approach of my research with the need for accounting, when appropriate, for the variety of individual positions.

Let me anticipate a historical remark about graduation theses which sheds light on the scope of my methodology. Graduation theses were written by the regents and are the most reliable source of information about the philosophy of the Scottish universities. Yet, it would be inaccurate to regard them as anything different from the summary of the curriculum of the Faculties of Arts. This means that graduation theses are not the product of a conscious search for innovation and personal research from the side of the regents. Rather, graduation theses enlighten us on the social dimension of the teaching of philosophy at the undergraduate level. This means that their philosophy is not the philosophy of the community of professional philosophers, it is rather the philosophy of the "laymen", it is the philosophy of students who, for the most part, did not pursue an academic career or personal research. Therefore, we can truly speak of a 'social dimension' of the philosophy of the graduation theses. I believe that this is the historical relevance of the graduation theses. The evidence of a common philosophy in the graduation theses is then even more important and revealing of the spirit of the time.

I shall now turn to a brief description of my primary sources, to the analysis of the academic context of the graduation theses and, in conclusion, to the relation between my research and the history of Scholastic philosophy.

2. *Theses philosophicae*: type of text and historical background[3]

The primary sources of my thesis are the graduation theses of the Faculties of Arts of the Scottish universities of the seventeenth century: King's and Marischal College in Aberdeen, St Salvator's and St Leonard's Colleges in St Andrews, and the universities of Edinburgh and Glasgow.

Graduation theses are texts of variable length (from 8 pages up to 60 pages), in quarto, written in neo-Latin and usually printed by the local printer, with the exception of St Andrews theses. Graduation theses were 'class theses', written by the regents for the whole class of students for the purpose of the graduation ceremony.[4] The regent is the lecturer in charge of the four-year curriculum of the Faculty of Arts: in the tradition of the regenting system, adopted by the Scottish universities in the seventeenth century, the same lecturer would guide his students through the learning of the four main branches of philosophy: logic, metaphysics, ethics and natural philosophy.[5] The curriculum culminated in a public graduation ceremony, to be held before the local community: this would include civil and religious authorities as well as other students and the regents. The candidate would engage in philosophical debate on a vast number of doctrines, to show his philosophical as well as rhetorical skills. This practice closely resembles the medieval disputationes, which were an important part of both the teaching and the examining of students.

[3] I shall present here the historical information required in order to understand the academic background of the graduation theses; I do not wish to offer an analysis of the universities in the first half of the seventeenth century. This is partly the object of broader studies, among others: Christine M. King (later Shepherd), *Philosophy and Science in the Arts Curriculum of the Scottish Universities in the 17th century*, Ph.D. Thesis, University of Edinburgh, 1974, and Steven J. Reid, *Humanism and Calvinism: Andrew Melville and the Universities of Scotland 1560-1625*, Farnham - Burlington (VT), Ashgate 2011. Important research on the history of the universities and of the graduation theses was inaugurated by P. J. Anderson, librarian of the University of Aberdeen between the late nineteenth and early twentieth century. I take the information regarding graduation theses from these works in particular: P. J. Anderson, *The Arts Curriculum*, Aberdeen 1892; and *Notes on Academical Theses*, Aberdeen, Aberdeen University Press, 1912; R. G. Cant, *The Scottish University in the XVIIth Century*, in Aberdeen University Review, 43 (1970), pp. 223-233.

[4] To my knowledge, the only exceptions to this practice are the individual sets of theses discussed by T. Mierbek, *Theses physicae de generatione et corruptione quas defendere conabor sub praesidio J. Echlini*, Edinburgh 1600; and by S. Decanus, *Positiones nonnullae physiologicae...sub praesidio M. Patricii Gordonii*, Aberdeen 1643. The candidates were granted the possibility to follow the graduation practice of their native country. Individual graduation theses were not prohibited, as we read for example in the *Fasti Aberdonenses*, Aberdeen, Spalding Club, 1854, in the *Leges Collegii Regii Aberdonensis*, p. 329: *"It is lykwayes speciallie ordered that ther be no privat lawreatione in aither of the tuo colledges, without consent of the earle Marischall, rector, principall, and regents of his colledge."*

[5] The regenting system was in use in the Scottish universities, albeit intermittently: C. M. King, *Philosophy and Science*, pp. 18-24. As the dates and authors of the graduation theses confirm, Edinburgh and St Andrews constantly applied the regenting system, while, for example, Aberdeen preferred the professorial system from around the 1620s to the 1640s: Sibbald, Seton and Leech were not regents, but professors of Natural philosophy in charge of the final year of the curriculum. Therefore, they wrote graduation theses more often than once every four years. In Glasgow the regenting system was reintroduced in 1642, when Dalrymple was appointed regent.

In order to understand the nature of the graduation theses, it is important to bear in mind that they were written for the purpose of the graduation ceremony. Class graduation theses do not belong to any traditional type of philosophical text, and can be said to be a genre on their own. Some theses are structured as short treatises, some others as short commentaries on Aristotle, yet it is evident that they are all examples of the broader category of graduation theses, which includes the oral discussion. There are no extant records of any actual discussion, but some graduation theses present lists of questions, usually under the heading 'Problemata', which can give us an idea of the sorts of topics discussed. A typical discussion would cover all areas of philosophy. The main scope of the graduation theses was to provide students with a summary of the curriculum, in which they would find what the regent considered to be the central doctrines and the best answers to philosophical debates. In some rare cases, the regent explicitly refers in the graduation theses to the candidates' discussion, and leaves the answers to them.

Given the variety of graduation theses and their broad spectrum, it is no surprise that the analysis can be detailed and long in some cases, and sketchy and incomplete in other cases, either within the same set of theses or between different sets. It takes a longer analysis to reveal where the main interests of the regent lay. Some sets are particularly detailed, for example the graduation theses by the three main regents of Edinburgh university in the 1610s and 1620s, Fairley, King and Reid. One regent who explicitly favours a shorter style in writing is for example Alexander Lunan, King's College, whose graduation theses of 1622 are a collection of short and often unexplained statements. We can generally claim that in the 1640s graduation theses are much shorter and more in the style of a handout than in the previous decades.

An indirect sign of the importance of the oral dimension is that the practice of printing graduation theses of the Faculty of Arts was started in Edinburgh only in 1596, to be quickly adopted by the other universities, but the graduation theses were not written with publication in mind. Graduation theses were discussed, and probably circulated as manuscripts among students before 1596 as well.[6] This means that, in general, a good deal of effort is required from the reader in order to understand the philosophy they expound. More often than not, graduation theses just present a brief explanation for extremely complex theories and, even in longer ones, the discussion of a particular theory cannot match the extensive analysis characteristic of Scholastic texts.

[6] We have evidence of a manuscript version of graduation theses: *Theses aliquot logicae, ethicae, physicae, et astronomicae, in publicam Disputationem exibendae, quas Adolescentes nonnulli Salvatorianae Academicae alumni jam laurea donandi, Praeside Jacobo Gleg, conabuntur* συν θειõ *propugnare*, St Andrews 1609. Ms 125 in Worcester College library, Oxford. This manuscript is the only extant version of this set of graduation theses.

I believe that graduation theses could be regarded as a shorter version of the Scholastic textbooks which became so popular in the late sixteenth and early seventeenth centuries. There are in fact similarities between Scholastic textbooks and graduation theses: both avoid the structure of quaestiones and articula typical of earlier Scholastic texts, both engage mainly with the clearest and best formulation of a problem, avoiding the longer process of obiectio and sed contra for each theory, and both abandon the practice of the commentary, as I mentioned above, with rare exceptions in the theses. Differences are no less evident: graduation theses are not written as an effective and exhaustive analysis of a branch of philosophy, or as an introduction to it, nor as a text to be used for teaching. In these similarities with Scholastic textbooks, graduation theses show the influence of the developments of Scholasticism in the early modern age.[7]

Graduation theses are thus the culmination of the undergraduate teaching in the Faculty of Arts. They stand in close relation with teaching, since they are a sort of summary of the four-year curriculum. Regarding the Scottish universities, there is evidence of the practice of teaching also in the form of course notes. They were usually compiled by students from lectures, and sometimes approved of by the regents. The most evident use for these notes is their circulation and sale among students as textbooks, since there is evidence that universities adopted the same notes over a period of several years.

The investigation of this material could shed light on university teaching. I have not used course notes alongside graduation theses in this thesis for the following reasons: 1) graduation theses are a more reliable source of information since they were written by the regents, while course notes are usually the result of the initiative of students; 2) graduation theses are the official philosophical production of the universities, while course notes are unofficial internal productions; 3) unlike course notes, the graduation theses are texts in which the regent was free to engage with philosophical debate without the needs imposed by teaching: graduation theses reveal much more of the personal philosophy of the regent; 4) alongside the issue of the chronological unity of my research, there is the fact of the strong unity of my primary sources. Graduation theses then seem to be the most historically reliable source of information about the philosophy of the universities.[8]

Graduation theses are usually divided into four sections: *Theses logicae*, *Theses ethicae*, *Theses physicae* and *Theses metaphysicae*. This division reflects the afore-mentioned quadripartition of the curriculum. This division is not always respected, and some differences between universities are well exemplified by changes in the structure of the

[7] For an analysis of the Scholastic textbook: P. Reif, *The Textbook Tradition in Natural Philosophy*, in Journal of the History of Ideas, 30 (1969), No. 1, pp. 17-32.

[8] For the analysis of coursenotes, see King, *Philosophy and Science*, *passim*.

theses: for example, this quadripartition applies well to both Aberdeen colleges, King's and Marischal, while Edinburgh and, less evidently, St Salvator's and St Leonard's in St Andrews tend to have *Theses astronomicae* in place of *Theses metaphysicae*; the former are virtually absent from Aberdeen theses. Sections on the distinctions between branches of philosophy (*Theses generales* or *Theses de disciplinis*) might be present, as well as *Theses geometricae* or *Theses mathematicae*. The invariable core of graduation theses is the tripartition into logic, ethics and natural philosophy.[9] Around these three sections, the regents were free to add a fourth or even a fifth section dealing with astronomy, geometry or metaphysics, which arguably reveals more of the specific character of each university. Thus, Aberdeen and St Andrews seem to have a greater interest in metaphysics than Edinburgh, which prefers to give space to science.[10] If we limit the investigation to natural philosophy, differences in structure among universities are less evident than in the graduation theses as a whole, and we can recognize a coherent and uniform natural philosophy.

Graduation theses derive their name from the practice of dividing the text into several theses that the regent proposes for the candidates' analysis. This division can be either into main theses and clarificatory sub-theses or simply into different theses. Here are two examples:

> I. Εσιν ἡ ὕλη δυναμις, το δ'ειδος ἐντελὲχεια. [*sic*] 2. de An. Tex. 2. 8. Met. 15.
> APPEND. I. Materia ergo essentialiter est potentia, eaque pura.
> 2. Per seipsam est receptiva formae, non per accidens superadditum. [Reid 1614, TP I]
>
> I. Philosophia speculatrix circa res versatur necessarias, a materia (quae erroris omnis, omnis

[9] The general title of graduation theses is usually *'Theses philosophicae'* or *'Theses logicae, ethicae, physicae et metaphysicae'*. I will refer to graduation theses, in general, as *Theses philosophicae*. Regarding the sections of the theses, I shall adopt the following abbreviations: TL for *Theses logicae*, TP for *Theses physicae* and TM for *Theses metaphysicae*, followed by the number of the thesis and, if necessary, by the number of the sub-thesis. I shall adopt these abbreviations even for those sets of graduation theses with different titles, very common in particular at King's College, Aberdeen: for example, W. Forbes, *Positiones aliquot logicae, ethicae, physicae, metaphysicae, sphaericae*, Aberdeen 1623; and for those sets whose physical theses are divided in sections, as for Sibbald 1625 for example. Therefore, each logical, physical and metaphysical section of the graduation theses will be referred to as respectively TL, TP and TM.

[10] By 'science' I mean astronomy, geometry, mathematics, not natural philosophy. As further evidence for this claim: Aberdeen theses are the most insightful in metaphysics, ranging from the discussion of the difference between essence and existence to the status of created beings in relation to God; St Andrews theses seem to develop a metaphysics of separate substances which anticipates later aspects of modern philosophy; Edinburgh theses basically offer the whole philosophy of science of the Scottish universities; finally, there are no extant theses for Glasgow, with the exception of a truly interesting set of theses from 1646, authored by James Dalrymple (later Viscount Stair).

> obscuritatis est latebra, et in cognoscendo difficultatis prima radix) mentis cogitationes avulsas [...]
>
> II. Mathematica non datur communis, ipse tamen vel maxime scientiae digna nomine [...] [Schewer 1614, TP I-II]

These passages are the first lines of the *Theses physicae* by Reid 1614 and Schewer 1614, regents respectively at Edinburgh and St Salvator's. Reid 1614 structures his set of theses as a commentary on Aristotle's passages, usually with the text in Greek as the main theses. The regent then moves on to expound Aristotle by means of several sub-theses which can lead to either the approval or the rejection of the main theses. The division between theses and sub-theses is particularly suitable for Reid's emphasis on commentary of the original text by Aristotle. Schewer 1614 structures his theses differently: there are no main theses and sub-theses, and each thesis is independent from others.

There are also other ways of structuring the theses. Some regents include either short treatises (as in the case of St Salvator's and St Leonard's 1629), or structure the theses in distinct sections, arranged by theses, as in Sibbald 1625[11], who divides his *Theses physicae* in: 1) *De pluralitate formarum in eodem composito*; 2) *An materia coeli sit diversa a materia sublunarium*; 3) *A quo coeli moveantur*; 4) *De speciebus intelligibilibus*; 5) *De praestantia intellectus et voluntatis*; 6) *A quo voluntas determinatur*. It appears that in this division by topics the regent focuses more on particular doctrines and less on covering the whole of natural philosophy.

We have seen that the practice of printing graduation theses was established in Edinburgh in the late sixteenth century, the oldest set of graduation theses available being J. Robertson's *Theses philosophicae*, Edinburgh 1596. Regarding the survey of extant graduation theses, one preliminary consideration is important. We can only speculate about the number of copies printed in the universities for the graduation ceremony. Considering an average number of twenty students per class, we can argue that perhaps twice as many copies were printed: one copy per student and the remaining copies distributed among the audience of the ceremony. There is, however, no record of this. This estimate and the consideration that graduation theses were not printed in order to be published and sold

---

[11] James Sibbald (1595-1647), minister of the Church of Scotland, regent and member of the Aberdeen Doctors. Graduated in 1618 at Marischal College, Aberdeen. First Professor of Natural Philosophy, appointed in 1620 as regent of the magistrand class. He wrote three graduation theses for the years 1623, 1625 and 1626. His theses cover the four-year curriculum, even if his teaching was restricted to natural philosophy. As the other Aberdeen Doctors, Sibbald was opposed to the National Covenant, and he eventually left Scotland for good in 1640. Died in Dublin in 1647. *FAM*, p. 33.

justifies the claim that graduation theses were, and are today, among *libri rarissimi*.[12] It is then rather surprising that a good number of graduation theses are still extant. The only graduation theses which did not survive are from the University of Glasgow, with the remarkable exception of James Dalrymple's 1646 theses. Here is a list of the existing theses for the other four universities:[13]

Aberdeen, King's College: A. Lunan 1622; W. Forbes 1623; J. Forbes 1624; W. Lesley 1625; J. Lundie 1626; J. Lundie 1627; A. Strachan 1629; A. Strachan 1631; D. Leech 1633; D. Leech 1634; D. Leech 1635; D. Leech 1636; D. Leech 1637; D. Leech 1638; P. Gordon 1643.

Aberdeen, Marischal College: A. Aedie 1616; J. Sibbald 1623; J. Sibbald 1625; J. Sibbald 1626; J. Seton 1627; J. Seton 1630; J. Seton 1631; J. Seton 1634; J. Seton 1637; J. Seton 1638; J. Ray 1643.

Edinburgh: J. Robertson 1596; W. Craig 1599; J. Adamson 1600; J. Knox 1601; J. Adamson 1604; J. Knox 1605; A. Young 1607; J. Reid 1610; W. King 1612; A Young 1613; J. Reid 1614; J, Fairley 1615; W. King 1616; A. Young 1617; J. Reid 1618; J. Fairley 1619; W. King 1620; A. Young 1621; J. Reid 1622; J. Fairley 1623; W. King 1624; A. Stevenson 1625; J. Reid 1626; R. Rankine 1627; W. King 1628; A. Stevenson 1629; J. Brown 1630; R. Rankine 1631; A. Hepburn 1632; D. Forrester 1641; T. Craufurd 1642; J. Wiseman 1643; D. Forrester 1645; T. Craufurd 1646; J. Wiseman 1647; D. Forrester 1649.

St Andrews:[14] J. Petrey 1603 (StS); D. Wilkie 1603 (StL); W. Wedderburn 1608 (StS); Anon 1608 (StL); J. Cleg 1609 (StS);[15] D. Robertson 1610 (StS); P. Bruce 1610 (StL); A. Henderson 1611 (StS); J. Strang 1611 (StL); J. Blair 1612 (StS); J. Wemys 1612 (StL); W.

---

[12] J. F. Kellas Johnstone, *Notes on Academic Theses of Scotland*, in Records of the Glasgow Bibliographical Society, 8 (1930), pp. 81-98: *"Arts Graduation theses are very rare, many rank among* "libri rarissimi". I take the estimate of the number of students from the dedicatory letter of each set of theses, which includes a list of the candidates. See also King, *Philosophy and Science*, Appendix 4, pp. 398 ff.

[13] Christine M. King based her research on the list in Harry G. Aldis, *A list of books printed in Scotland before 1700*, Edinburgh, National Library of Scotland, 1904, reprinted 1970. A more complete list is in Alfred W. Pollard, *A short-title catalogue of books printed in England, Scotland & Ireland and of English books printed abroad, 1475-1640*, 2nd ed. revised and enlarged by W. A. Jackson - F. S. Ferguson, London, Bibliographical Society, 1976-1991. The main difference between Aldis and Pollard is the Clarke's collection, a bundle of graduation theses from St Andrews, missing from Aldis, now in Worcester College library, Oxford. I am grateful to Joanna Parker, librarian of the Special Collections, for the identification of these St Andrews theses as the Clarke's collection.

[14] StS stands for St Salvator's, StL for St Leonard's.

[15] Ms at Worcester College library, Oxford. See above, p. 14, footnote 3.

Lamb 1613 (StS); W. MacDowell 1613 (StL); J. Schewer 1614 (StS); A. Bruce 1614 (StL); D. Monroe 1615 (StS); Anon. 1615 (StL);[16] J. Wemys 1616 (StL); R. Baron 1617 (StS); J. Carr 1617 (StL); W. Martin 1618 (StS); A. Bruce 1618 (StL); R. Baron 1621 (StS); J. Baron 1627 (StS); A. Monroe 1628 (StS); M. Murray 1628 (StL); J. Ramsey 1629 (StS); J. Wedderburn 1629 (StL); J. Mercer 1630 (StL); J. Barclay 1631 (StS); W. Wemys 1631 (StL); A. Monroe 1632 (StS); J. Mercer 1632 (StL); M. Murray 1634 (StL); J. Armour 1635 (StS); W. Wemys 1635 (StL); J. Wood 1637 (StS), D. Nevaius, 1648 (StL).[17]

The general picture is that we are in possession of an almost complete list for Aberdeen, Edinburgh and St Andrews, with the regrettable loss of almost all Glasgow theses. As far as I know, the list until 1649 I provide here, a combination of Aldis, Pollard and personal research, is the most complete.

3. Protestant Scholasticism

In section 1 of this introduction I sketched a proof that the graduation theses are part of the tradition of Protestant or Reformed Scholasticism. This notion is indeed called into question by scholars for two different yet converging reasons: on the one side, Catholic scholars tend to restrict the notion of Scholasticism to 'Catholic Scholasticism', and, even more precisely, to Thomism as the appropriate style in philosophy for a Catholic philosopher. On the other side, non-Catholic scholars tend to mark the difference between the Roman Church and the Reformed churches in terms of the rejection of Scholasticism tout court, which allegedly took place because of the Reformation. It is hard not to detect a political agenda behind these two positions, both unmindful of two pertinent considerations: 1) Scholasticism was not exclusively adopted by Catholic philosophers; 2) Scholasticism was not ended by the Reformation, and, for example in Scotland, after the conversion of the Scottish church in 1560, flourished well into the seventeenth century.

---

[16] King did not find any records for the regents at St Leonard's for the years 1603-10 and 1614-16. The unique copies of the theses for 1608 StL and 1615 StL that I have read at Worcester College library have no title page. The names of the regents are unknown.

[17] The Clarke's collection includes the following theses, some of which are unique: 1609 StS, 1610 StS and StL, 1611 StS and StL, 1612 StS, 1613 StS and StL, 1614 StS, 1615 StS and StL, 1616 StS and StL, 1617 StS and StL, 1618 StS. J. Wood 1637 StS is not listed in Pollard, and the following theses from the Clarke's collection are not included in Shepherd 1974: 1610 StS and StL, 1611 StS, 1612 StS, 1613 StL, 1614 StS, 1615 StS and StL, 1616 StS and StL, 1617 StS, 1618 StS.

Now, with regards to the notion of 'Reformed Scholasticism', there are two main questions to be asked: 1) what is the actual state of research? and, 2) is there a criterion in virtue of which it is possible to define a tradition in Scholasticism as 'Reformed' or 'Protestant'? I shall seek to analyse the present state of research on Reformed Scholasticism and argue that Scottish Scholasticism can shed important light on the whole notion, and I shall explain why I believe it possible to identify Reformed features in the natural philosophy of the theses.

3.1 The *Theses philosophicae* and the historiography of Scholasticism

Reformed Scholasticism is still underexplored territory. Scholasticism in general deserves more attention than scholars have been willing to show.[18] It is a merit of the Catholic universities to have fostered the interest in Scholasticism, in particular following the Encyclical Letter *Aeterni Patris* of 1879, in which Pope Leo XIII officially adopted Thomistic philosophy for the teaching of the Roman Church. The work of Etienne Gilson is probably the highest and among the first examples of this renewed interest in Scholasticism. But by no means is it the only one: the works of P. O. Kristeller, C. Schmitt, B. P. Copenhaver among others[19] mark the beginning of a better understanding of Renaissance and early modern philosophy, which includes the Scholastic and Aristotelian traditions. Even if research is currently under way,[20] the variety and depth of Scholasticism

[18] M. W. F. Stone, in *The Cambridge Companion to Early Modern Philosophy*, D. Rutherford (ed.), Cambridge, Cambridge University Press, 2006, pp. 299-327, in particular pp. 302-304 and 317-320 addresses the role of Protestant Scholasticism: *"Protestant Scholasticism made an important contribution to the theology and philosophy of the period"* (p. 302), in particular in relation to great figures of early modern philosophy, such as Leibniz, Locke and Kant. The author underlines the fact that the interpretation of Protestant Scholasticism as *"a period of intellectual decline"* has been put forward by twentieth-century Protestant theologians with no interest in Scholasticism (p. 317). An invaluable work for the reassessment of Protestant Scholasticism is R. A. Muller's *Post-Reformation Reformed Dogmatics*, Grand Rapids (MI), Baker Book House, 1987. Muller connects the development of Protestant theology with Scholasticism, and even if his interest mainly lies in the theological aspect, his work can shed light on the history of philosophy as well.

[19] Among other works: C. B. Schmitt, *Aristotle and the Renaissance*, Cambridge (MA), Harvard University Press, 1983; *John Case and the Aristotelianism in Renaissance England*, Kingston, McGill-Queen's University Press, 1983; *The Aristotelian Tradition and Renaissance Universities*, London, Variorum Reprints, 1984; *Reappraisals in Renaissance thought*, edited by C. Webster, London, Variorum Reprints, 1989. P. O. Kristeller, *Renaissance Thought and the Arts*, Princeton - Oxford, Princeton University Press, 1990. B. P. Copenhaver - C. B. Schmitt, *Renaissance Philosophy*, Oxford, Oxford University Press, 1992. These authors contributed in a decisive way to the idea that Renaissance philosophy was an autonomous area in the history of philosophy. They caused a shift in the scholarly opinion on Renaissance Aristotelianism and Scholasticism, contributing to the understanding of the many different aspects of Renaissance philosophy, and provided guidelines for future research in many areas: from the history of the universities to the relations with early modern philosophy.

[20] J. Schmutz, *Bulletin de scholastique moderne*, in Revue Thomiste, 100 (2000), No. 1, pp. 270-341. The author offers an insightful review of recent publications on Scholasticism and also proposes lines of

requires much more investigation in three main directions: 1) the relation between Scholasticism and Renaissance philosophy; 2) the relation between Scholasticism and early modern philosophy - the best known area, in virtue of the attempts to understand the background of Cartesian philosophy; and finally 3) the relation between Catholic and Protestant Scholasticism. My interest lies in shedding light on point 3: I shall also hint at possible lines of research regarding point 2.

Prior to the acceptance of the notion of a 'Reformed Scholasticism', scholars debated the definition of Scholasticism in the early modern era. Whereas there seems to be no doubt that Scholasticism in the early modern era is a distinct philosophical movement from the Scholasticism of the Middle Ages, there is no agreement regarding what makes it a distinct movement. The most common formulae are 'second Scholasticism', 'modern Scholasticism', 'late Scholasticism', 'academic Scholasticism', 'Renaissance Scholasticism', 'Baroque Scholasticism', which exemplify well the extent of the disagreement among scholars.[21] The premise of theses formulae seems to be the assumption that, despite the differences among the schools, Scholasticism in the early modern era was ultimately a unitary movement. I argue that this disagreement could be resolved by appealing to a different criterion of classification, philosophical rather than historical. If it is true that some degree of unity within Scholasticism in the early modern era is evident, nevertheless the division into 'Catholic Scholasticism' and 'Protestant Scholasticism', already in use in the history of theology, might be profitable in history of philosophy as well.

The definition of a historical period reveals the point of view of the historian, just as much as it reveals characteristics of the defined object. If we accept the idea that a definition in the history of philosophy cannot exhaustively define its object, then I believe that a division of Scholasticism on the basis of the faith of the philosophers can be a useful one. Clearly, all the afore-mentioned definitions shed light on some aspects of the Scholasticism of the modern age. Yet, the question raised in my research seems to regard

---

research. The analysis seems a little unbalanced in favour of Spanish Scholasticism. Historically more interesting is M. Forlivesi's introduction *A Man, an Age, a Book* to the volume *Rem in seipsa cernere: Saggi sul pensiero filosofico di Bartolomeo Mastri (1602-1673)*, M. Forlivesi (ed.), Padua 2006. Forlivesi engages with the account of the Scholastic tradition in the period from the late middle ages to seventeenth century and offers extensive bibliography on the subject. Once again, the focus is on Catholic Scholasticism. Scotland is not mentioned and the sole reference to England is made for Britain, page 48. Despite this, Forlivesi's analysis is a most accurate account of our current knowledge of Scholasticism. Some other fundamental texts in Renaissance philosophy are: N. Kretzmann - A. Kenny - J. Pinborg (eds.), *The Cambridge History of Later Medieval Philosophy*, Cambridge, Cambridge University Press, 1982; C. B. Schmitt - Q. Skinner - E. Kessler (eds.), *The Cambridge History of Renaissance Philosophy*, Cambridge, Cambridge University Press, 1988; J. Hankins (ed.), *The Cambridge Companion to Renaissance Philosophy*, Cambridge, Cambridge University Press, 2007.

[21] Forlivesi, *ivi*, pp. 106-114. Forlivesi seems to approve of 'Renaissance Scholasticism', while pointing out that all the formulae are in some sense profitable (p. 112).

more what is specific to the Scottish Scholasticism as a form of Scholasticism, rather than what Scholasticism is in general.

The advantage of the definition that I set out to employ is that it accounts well for the philosophy of the graduation theses. In fact, we shall see that graduation theses put forward a 'Reformed' philosophy not simply on the general recognition that the regents were philosophers who "happened to belong" to a Reformed faith community; rather, the graduation theses expound doctrines whose philosophical character originates from a Reformed confession of faith. My two key examples will be the theory of the relation between accident and substance (part I, chapter 4) and the rejection of natural theology (part II, chapter 3).

The formula 'Reformed Scholasticism' and the application of this formula to the whole of Scholasticism prompt the not easily answerable question of the acceptability of a theological category in a philosophical categorisation. There is no doubt that the division into Reformed and Catholic philosophy is primarily motivated by religious events and theological doctrines, and that the very reference to theological doctrines in a philosophical context might seem an illicit move. Yet, I believe that this criticism can be rebutted in two ways:

1) the graduation theses do not openly engage with theology, because the Faculties of Arts were dedicated to the teaching of philosophy. There is then among the regents the awareness that philosophy is a distinct discipline from theology. I shall argue that the natural philosophy of the theses shows this attitude well. Yet, the regents feel compelled to investigate theological doctrines insofar as they have consequences for philosophy. Even if this investigation is conducted within the limits of and according to the principles of philosophy, nonetheless it is prompted by theological doctrines. Within Scholasticism, the specific character of the graduation theses is exemplified by such philosophical theories: the regents understood themselves to be different as philosophers from Catholic Scholastics primarily in virtue of the rejection of the Catholic theory of accidents, grounded in their Reformed reading of the Eucharist as a symbol.

2) More generally, the importance of religion and theology in the philosophical debate of the seventeenth century should not be underestimated. Later, religion is thought of as a private aspect of men's lives, but in seventeenth-century Europe the public dimension of religion was very prominent, and an important part of the struggle of the new philosophy and science was for independence from religion and theology. I believe that the graduation theses are fully Scholastic in spirit when it comes to the relation between philosophy and theology; yet, a degree of autonomy of philosophy from theology was part of the

Scholastic tradition as well. As we will see, even if graduation theses do not engage with theology, they are nonetheless influenced by religion.

In sum, the graduation theses contain, explicitly or otherwise, natural philosophical theories which are motivated by the faith of the regents. Religion is then an acceptable basis for distinctions within Scholasticism.[22]

Other formulae in use are those of 'early modern philosophy' and 'modern philosophy'. These formulae seem to include, especially in the analytic tradition, almost exclusively the philosophy after Descartes. This approach tends to exclude philosophical traditions such as Renaissance or Scholastic philosophy of the sixteenth and seventeenth centuries. Even if Scholasticism is an important part of the narrative of early modern philosophy (as the increasing literature on the Scholastic background of Cartesian philosophy shows), I shall employ 'early modern philosophy' and 'modern philosophy' to refer only to the Cartesian and post-Cartesian traditions. In a sense, this choice is motivated by the use in the graduation theses of the expressions *'Scholastici'* and *'Moderni'* as referring respectively to those philosophers who follow the philosophy of the schools and those who do not. Therefore, we are confronted with three traditions in the analysis of the theses: primarily 1) Reformed Scholasticism and 2) Catholic Scholasticism; then 3) early modern philosophy. Renaissance philosophy, in particular in the form of Humanism, is in secondary position.

3.2 The doctrine of the Fall: a religious premise to natural philosophy

The doctrine of the Fall is part of the Christian faith. Historically, it gained greater importance because of the Reformation: Reformed theologians, philosophers and laymen felt the corrupt condition of human nature in a more vivid way than Catholics. This doctrine finds its way into some graduation theses, and into natural philosophy more often than into moral philosophy, where its importance should be more evidently perceived. A corrupt state entails our essential incapacity for good moral behaviour.

In the graduation theses, the doctrine of the Fall is exploited as a premise to natural philosophy, and it seems to imply that not just the moral judgment of men is impaired, but

[22] My focus is on natural philosophy only. Perhaps surprisingly, the Reformed Scholastic character of the theses is best exemplified by natural philosophical theories rather than by, for example, the theory of free will. In fact, even if Scottish Reformed regents did not believe in free will, graduation theses expound the doctrine of free will because it is the best possible solution to the question of human action according to the principles of human reason alone. On the one side, this evidence shows the degree of autonomy that philosophy was granted in the Scottish universities; on the other, it is even more remarkable that in some natural philosophical theories, rather than in moral ones, the regents perceived 'good philosophy' to be in harmony with their religion.

the understanding of the natural world as well. Therefore, corruption affects human reason in both its moral and theoretical aspects, and therefore in respect of both will and intellect.

> Lapsu flebili, non modo paralysi dissoluti affectus, transversum acta voluntas, sed et Thebanis sphingibus, Cymmerijs tenebris obtenebrata mens.
> Lugubris conditio humana non modo disciplinae practicae medelam, sed et scientiae contemplativae collyrium et solem requisivit. [Robertson 1596, TP 1][23]
>
> Execrabili hominis Apostasia, sicuti vitiati sunt affectus, corrupta ac depravata voluntas: ita mens densissima ignorantiae caligine obnubilata est.
> Morborum animi, cujus medicina est Philosophia [...] [King 1612, TP I]

These two passages from Robertson and King claim that the corruption due to the Fall is not limited to the will, but extends to the intellect as well. Robertson talks of the 'grievous human condition', while King talks of the 'diseases of the human soul'. In both passages, the remedy for this condition is contemplative science or philosophy.

This picture applies particularly well to our understanding of prime matter, whose analysis follows each of these two passages. In fact, as we will see, prime matter is most obscure to us, because it is not endowed with form. Yet, the prominence given to the doctrine of the Fall by Robertson and King is remarkable: natural philosophy as a whole should be regarded as an enterprise originally impaired by the limitedness and corrupt state of our understanding, which originated with the Fall. Philosophy is a remedy, but it does not to seem to be a solution.

Now, the reference to the doctrine of the Fall in the context of natural philosophy seems to be a consequence of the Reformed religion of the regents, in this case identifiable as a form of Calvinism. Can we say that this reference is sufficient ground for the definition of the Scottish Scholasticism as 'Reformed Scholasticism'? I think it is not. In fact, the doctrine of the Fall does not affect the philosophy of the regents. More precisely, there appears to be no philosophical doctrine which is different from an equivalent doctrine in a Catholic Scholastic context because of the doctrine of the Fall. I do not wish to underestimate the importance of the doctrine of the Fall in shaping the worldview of the regents; yet, this doctrine seems to qualify as a religious premise rather than a philosophical theory. Therefore, the reference to the Fall should be understood as a sign of

---

[23] A translation of the *Theses physicae* of Robertson 1596 is in the Appendix.

the Reformed religion of the regents, rather than an aspect of their religion which actively shapes their philosophical argumentations.[24] I have argued above that this is not the case for the Reformed reading of the Eucharist, and the belief in the Calvinist *sensus divinitatis*: these religious doctrines respectively shape the regents' theory of the relation between accident and substance, and ground the rejection of natural theology. In these cases in fact, the regents oppose Catholic Scholasticism on the basis of their religion, and bring about fundamental changes with respect to Catholic Scholastic philosophical theories.

The investigation of graduation theses can prove an extremely important step to a better understanding of Scholasticism. The characteristic of the graduation theses as the official philosophical production of the Scottish universities enables the historian of philosophy to investigate a coherent and unitary corpus of Scholastic texts. It is evident that Scholasticism in the early modern era was an incredibly variegated philosophy, with differences on the basis of nationality, religion, philosophical heritage and political pressure. Scotland is a particularly suitable territory for the investigation of academic Scholasticism, a territory in which the national element coheres with a philosophical unity.

With regards to the Reformed aspect of Scholasticism, the graduation theses are a form of Reformed Scholasticism. The advantages of the graduation theses that I have pointed out with respect to Scholasticism in general are not less important in the context of the investigation of 'Reformed' Scholasticism. In particular, the graduation theses are purely philosophical texts, which can help to qualify Reformed philosophy without references to Reformed theology and Reformed theologians, though it is these references that have dominated approaches so far.[25]

---

[24] It is an established interpretation of the outcome of the Reformation that the worldview of the Reformed countries became increasingly favourable to a scientific research independent of religion, fostering the scientific revolution of the seventeenth century. P. Harrison, *The Bible, Protestantism and the rise of natural science*, Cambridge, Cambridge University Press, 1998, in particular chapter 2, believes that a direct consequence of the Reformation was the distinction of spheres between the two books, that of nature and that of revelation, thus benefiting the autonomy of natural philosophical research. Moreover, even the non-mediated access to the Scriptures, comparably greater in the Reformed countries than in the Catholic ones, favoured the spirit of independent research.

[25] C. R. Trueman - R. Scott Clark (eds.), *Protestant Scholasticism*, Carlisle, Paternoster Press, 1999; W. J. van Asselt - E. Dekker (eds.), *Reformation and Scholasticism*, Grand Rapids (MI), Baker Academic, 2001; W. J. van Asselt, *Protestant scholasticism: some methodological considerations in the study of its development*, in Dutch Review of Church History, 81 (2001), pp. 265-274; *Scholasticism Protestant and Catholic: Medieval sources and methods in seventeenth-century Reformed thought*, in J. Frishman - W. Otten - G. Rouwhorst (eds.), *Religious Identity and the Problem of Historical Foundation. The Foundational Character of Authoritative Sources in the History of Christianity and Judaism*, Leiden, Brill, 2004, pp. 457-470. E. Rummel, *The Humanist-Scholastic Debate in the Renaissance and*

4. Outline of the thesis

The thesis is divided into two parts. The first part is about prime matter, and consists of four chapters: 1) Materia prima: quod sit et quid sit; 2) De potentiis materiae primae; 3) De proprietatibus materiae primae; and 4) De Transubstantiatione. The investigation will show the Scotistic influence in the graduation theses and the coherence between Aristotelianism and Reformed religion in the theory of the relation between accidents and substance.

The second part is about movement, and consists of three chapters: 1) Motus: general features of movement; 2) The movement of *gravia* and *levia*; and 3) The movement of the heavens. The Scholastic theory of movement and the Reformed religion of the regents will have implications for the rejection of natural theology.

The Conclusions include the account of the reception of Aristotle in the theses: a Humanist renewed interest in the Greek text of Aristotle is conjoined with the Christian reading of Aristotle and the specific Reformed interpretation of Aristotle on the theory of substance.

In the Appendix I provide the translations of four sets of *Theses physicae*, extracts from Robertson 1596, Aedie 1616, Reid 1626 and Dalrymple 1646. These sets of theses are particularly interesting for the following reasons: 1) Robertson 1596 is the oldest set available to us; 2) Aedie 1616 is the oldest set for Aberdeen and it includes unique sections on special physics; 3) Reid 1626 and Dalrymple 1646 critically engage with the tradition of Scholasticism, in a way unknown to the other regents.[26]

Reformation, Cambridge (MA), Harvard University Press, 1995, focuses also on the Humanist counterpart of Reformed Scholasticism.

26 Theses texts (in particular Aedie 1616) show a variety of natural philosophical doctrines which can be difficult to contextualize for a contemporary reader. I have provided some references already, but I am planning to provide fuller references to them in later publications.

Part I, chapter 1

Materia prima: quod sit et quid sit

1. The relevance of prime matter in Scholastic natural philosophy

Prime matter (*materia prima*) is the stuff all bodies are made of. It is a common Scholastic theory that prime matter is the root of potentiality, the underlying principle on which form acts as the informing principle. The result of these two principles is a compound (*compositum*): a union of form and matter, a union which is essentially one because the two principles alone are not able to exist one without the other. Aristotle considers only form and matter principles *per se* of the compound, while calling privation (*privatio*) a principle *per accidens* of the compound, because 1) it is not a being in the full sense, since it is an absence of being; and 2) it is ultimately absence of form: therefore privation is reduced to form, because the absence of a (new) form is always the presence of a form (*Phys.* I).

Generally speaking, Scholastics claim that every body is a compound of matter and form. This is different from hylomorphism, which entails that all beings (with the exception of God) are made of form and matter, including, for example, angels. Not all Scholastics accept this theory, which is traditionally held by the Franciscan school. When it comes to natural philosophy, which deals with the realm of things-in-becoming, we can say that hylomorphism is shared by all Scholastics. Scottish regents also embraced hylomorphism.

Prime matter is then one of the two principles all natural things are made of: this is enough to show how important a notion it is. Alongside this, prime matter is the root of becoming. 'Becoming' (*fieri*) is the name given to any changes whatsoever: becoming is the continuous process of 'passing away-coming to be-passing away' which any compound undergoes in the course of nature. It was debated among Scholastics whether all becoming was included in the notion of movement or not. As it will appear in part II, the Scholastic notion of movement (*motus*) does not coincide with our contemporary notion, as it includes phenomena we would not call a 'movement' today. An even broader term is change (*mutatio*), which also includes changes which take place in an instant, and that some Scholastics and some regents tend to exclude from the number of 'movements'. A theoretical unity of all these processes is given by their common material cause, prime matter, the passive principle of the compound which causes (in the sense of 'material cause') the succession of forms, therefore the succession of beings.

In the structure of a compound, matter is on the side of potency and form on that of act. In general, this principle is accepted by Scholastics. In its strict version, it is famously a Thomistic doctrine. It is, however, not shared by all Scholastics, and the regents in general reject it. In natural philosophy, no specific contradiction between this doctrine and experience is evident, while regents felt compelled by philosophical arguments to go beyond this doctrine when dealing with prime matter under a metaphysical point of view. The notions of act and potency are most important in Scholastic philosophy. It may suffice here to define potency as: a) *"first, the principle of movement or change that we find in something else or in the same thing as something else"*; and b) *"the principle by which one thing is changed or moved by something else or by itself as other [from itself]"* (Aristotle, *Met.*, V, 12, 1019 a 15-20);[1] and act as *"a being which has some sort of actuality, thanks to which it is not nothing"* (Suárez, *DM*, 13, 5, 7.); and *"the 'existing' of one thing"* (Arist, *Met.*, IX, 6, 1048 a 32).[2] I believe that these definitions are general enough to serve as introductory definitions: we will see how the regents will employ them in their philosophy.

So, prime matter is: 1) the common material principle underlying all natural substances; 2) the root of becoming, by being the principle of receptivity of form; 3) the principle of unity in nature, both metaphysical and logical. This last aspect is particularly important to my work, as it also gives theoretical unity to part I. Following the metaphysical order within prime matter I have structured part I as follows: I first focus on the essence, secondly on the powers and thirdly on the properties of prime matter. The essential connection between prime matter and movement establishes a unity between parts I and II.

In this first chapter I shall investigate: 1) the evidence for the existence of prime matter (the Scholastic question 'utrum sit'), by means of three arguments: from natural becoming, by eminence and by negation; 2) the arguments for the definition of the essence of prime matter, or what prime matter is ('quid sit'). The answers given by the regents are that prime matter exists and that it is a 'receptive entitative act'.

---

[1] a) 'Δύναμις λέγεται ἡ μέν ἀρχὴ κινήσεως ἢ μεταβολῆς ἡ ἐν ἑτέρῳ ἢ ᾗ ἕτερον'; b) 'ἀρχὴ μεταβολῆς ἢ κινήσεως λέγεται δύναμις ἐν ἑτέρῳ ἢ ᾗ ἕτερον, ἡ δ' ὑφ' ἑτέρου ἢ ᾗ ἕτερον.' My translation.

[2] '[E]"στι δὴ ἐνέργεια τὸ ὑπάρχειν τὸ πρᾶγμα.' My translation.

2. Prime matter: quod sit

By this expression Scholastics mean 'that it is'. A proof 'quod sit' about prime matter aims to show 'that prime matter is', or, with a more contemporary terminology, 'that prime matter exists'. The Latin *sit* is philosophically more neutral than 'exists', since it only entails the attribution of being to a subject, while *existentia* is more precise. Existence *"dicatur esse modus quidam essentiae intrinsecus quo formaliter res dicitur esse actu sive extra suas causas"*, according to Eustachius a Sancto Paulo.[3] So, existence is a mode intrinsic to an essence by which we can say that an essence is in act outside of its causes. To say that something is and that something exists are then different claims. I will use the expression 'existence', also because the analysis of the *Theses philosophicae* will show precisely that according to the regents prime matter exists in the sense employed by Eustachius.

The claim that prime matter exists is different from the claim that matter exists: there is hardly any debate in Scholasticism over the existence of matter, while uncertainty about prime matter is strong. What is the difference? Matter is commonly intended as the matter of a given compound, and no doubt is possible regarding its existence: it is a fact that all bodies are also material. This is the Aristotelian notion of matter as potential principle of the compound. On the contrary, prime matter is the metaphysical notion of matter before information, a general, underlying principle of which we have no direct experience. The inference from *this* matter in a compound to prime matter *in general* is not immediate, and requires justification. Neither specific individual matters are species of the genus prime matter, so that the existence of the species entails the existence of the genus. Indeed, prime matter is not distinct from the individual matter and vice versa. No attribute of prime matter is withdrawn from informed matter, thus the genus-species parallelism does not work.

Inevitably though, any demonstration of the existence of prime matter has the existence of informed matter as a premise. There is more than a simple inference from informed matter to prime matter and regents deploy arguments in favour of the existence of it. They all agree that prime matter is and that it also exists in a more precise sense.

---

[3] *SPhQ*, IV, II, II, IV.

2.1 Argument from natural philosophy

Scholastics developed a range of demonstrations of the existence of prime matter which vary from more theological ones to metaphysical and physical ones. I wish to analyse the argument in its physical form because it is limited to the realm of natural philosophy, it aims at being self-sufficient with regard to other arguments, and it is favoured by the majority of the regents. The sets of theses offering the best formulation are Wedderburn and Ramsay 1629, a joint set of graduation theses for the students of the colleges of St Leonard's and St Salvator's in St Andrews. It relies on the principle that *in omni causarum genere datur aliqua prima causa* and runs as follows:

> Ducitur ex naturali rerum generatione: ex nihilo quicquam gigni non potest, ut experientia constat. Ergo, ex aliquo praeexistente, quod in re genita maneat. Id autem non est forma, ea namque denuo inducitur. Est igitur quidpiam, quod advenientem formam excipiat, et unum idemque permaneat, id vero est materia, quam primam dicimus. [Disputatio physica, an detur materia prima, et qualis ea sit]

If we couple the principle that in every causal genus a first cause is given with the principle that nothing can come from nothing, according to Wedderburn and Ramsay we are compelled to say that prime matter exists. Not simply matter, but prime matter: in fact, the matter of a compound is part of the premise of the argument, and a datum of our experience. The argument wants to bridge the distance between individual portions of matter we are aware of in our experience and prime matter by means of metaphysical principles. The structure of this argument is inevitable, given the ontological status of prime matter.

The relevance of this argument in natural philosophy is that it appeals to the causes of becoming in the natural world. Things become (come-to-be and cease-to-be): this whole process would be unintelligible if deprived of a metaphysical and physical unity, which is provided by prime matter. If prime matter did not exist as an underlying common principle of things in becoming, then 1) things would come out of nothing; or 2) things would be created and sustained continuously by God. In the first case we would have a contradiction: it is not possible that natural substances come from nothing, that is, from a material nothing, unless by means of creation. In the second case, we would have a continuous act of creation required to avoid the contradiction in the first case. If prime matter is posited, continuous creation is not necessary. In itself, the second case is not contradictory, it simply undermines the 'independence' of the natural world. According to the regents, the natural world is created, so ultimately dependent on God's causality in coming-to-be and in continuing-to-be; yet they are also aware of the independence of natural philosophy as a discipline and of the natural world as a realm on its own. Thus, the idea of a continuous creation is rejected. I believe that this approach is central in the theses and will surface again in my analysis.

This argument from natural becoming can be said to be Aristotelian in spirit, but less Aristotelian in letter. The Aristotelian side of it is the attention paid to the philosophical justification of change by the search for an unchanging principle; by the reconduction of plurality to unity; and finally by the rejection of absolute nothing as part of reality (for instance, *Phys.* I). What is not much Aristotelian is the very notion of prime matter: Aristotle never directly enquired into a prime matter with all the qualifications that Scholastics attributed to it. We might say that the notion of 'prime matter' has its full meaning only in the framework of a philosophical theology of created beings.

2.2 Other arguments: *per eminentiam* and *per negationem*

A physical argument is not the only way to prove the existence of prime matter. How arguments are structured reflects the sort of knowledge we can have of the *demonstratum*. In the previous case, in physical terms, we must deduce the existence of prime matter by means of metaphysical principles because our experience alone does not show that prime matter is. We simply do not *know* prime matter in the way we know natural substances because prime matter is not a part of our experience. This is due to the lack of form: we cannot say that our knowledge of things is limited per se to compounds of form and matter, but we can say that in nature we only experience compounds of form and matter. All knowledge of the physical world other than direct knowledge from experience must be obtained by philosophical means. This will be particularly important in the analysis of the heavens, in part II, chapter 3.

This specific status of prime matter is reflected in two other arguments employed by regents: they can be labelled the argument *per eminentiam* and the argument *per negationem*.

The first is found in Lunan 1622, who writes that prime matter:

> esse ens non ens, omnia nihil, existere non existere, potentiam non potentiam, actum non actum, unam

> multam, Singularem universalem, substantiam non substantiam, corpoream incorpoream, formatam informatam, quantam non quantam, omnia nihil appetere. [TP 1]

This passage is unique in all the theses for its explicitness. It is an inclusive list of all the oppositions available about prime matter. In this passage, Lunan employs the *via eminentia* to make us aware of the status of prime matter: usual oppositions derived from the terminology about the finite world do not apply to prime matter because it is before (and thus above) those determinations. Terms such as 'existent', 'singular', 'bodily', 'quantified', lose much of their original meaning when predicated of prime matter, which is essentially all of those determinations and at the same time exclusively none of them. In a strict sense, this is not an argument: Lunan does not proceed from premises to conclusions in order to prove the existence of prime matter. What he does is to show the non-natural status of prime matter and the attendant difficulties we experience when trying to define it. As it appears, this passage already implies the notion of what prime matter is, *quid sit*.

The second argument is taken again from St Andrews 1629:

> Per negationem, ita ut ab ea [prime matter] omnes perfectiones determinatas removeamus, dicendo: eam non esse substantiam, non quantitatem, non qualitatem, nec ullam ex determinatis entis speciebus. [...] Deinde docuit *Arist.* eam cognosci per analogiam: quemadmodum enim se habet aes ad statuam, cera ad sigillum recipiendum ita se habet materia prima ad formas recipiendas. [*ibidem*]

Wedderburn and Ramsay use much of the terminology we find in Lunan 1622 in a different context: where Lunan places prime matter above finite beings, Wedderburn and Ramsay on the contrary subtract qualifications from prime matter. The result is similar: prime matter is said not to have the perfections we find in the compounds of form and matter.

Via negativa and *via eminentiae* lead to the same conclusion when applied to prime matter; this does not mean that prime matter is in any way 'more perfect' than finite beings, because it is not. Prime matter is really deprived of perfections and it is left with the most basic perfection of being non-nothing. By eminence Lunan does not mean metaphysical eminence (an absolute perfection, which belongs to God) but some sort of epistemological "aboveness" of the notion of prime matter with respect to the notions of compounds.

3. Prime matter: quid sit

Prime matter exists: then, what is it? This question is addressed in many passages of the *Theses philosophicae*. The aim of this question is to find what the essence of prime matter is: what are, broadly speaking, its characteristics, once it has been established that it exists. This is the principal line of enquiry that the regents pursue with respect to prime matter and the question about the *quod sit* is merely preliminary to this enquiry. Nonetheless, answering the *quod sit* contributed to the clarification of some points which will be present in the discussion: 1) prime matter is not an object of our direct experience;[4] 2) it is not a substance like others, therefore the attributes of natural substances do not apply to it; 3) in establishing what prime matter is, the boundaries of natural philosophy are sometimes allowed to encroach on metaphysics.

The centrality of prime matter is such that the answer that regents give to what its essence is will have an influence on their natural philosophy as a whole: this point will be explicit in part I, chapter 4. The importance they attribute to the subject is also shown by the succession of topics: usually, the discussion of prime matter comes first in natural philosophy, for metaphysical reasons (it is a principle of all bodies) and for logical reasons (clarifying what prime matter is enables us to go further in the analysis of natural bodies).

Early modern Scholasticism as a whole inherited the doctrine of prime matter as pure potency (*pura potentia*) from medieval Scholasticism. It was famously endorsed in the thirteenth century by Thomas Aquinas, who claimed that prime matter is pure potency in a strict sense, and by John Duns Scotus, who reformulated the doctrine in a very influential way. Scotus denied the intelligibility of the notion of pure potency per se, and introduced an act proper to prime matter in order to avoid the contradiction of something existing yet existing as a pure potency, a pure possibility-to-be. The influence of Scotus's theory was enormous in Scholasticism, and considered by many as a definitive improvement in metaphysics. Usually, Thomists remain strong opponents of Scotus until today, even if

[4] King 1620, TM VIII, integrates this point by saying that *"intellectus noster tantum mensura est rerum artificialium"*. This is a Scholastic slogan. Our intellect is the measure of artificial things only, namely, things that our intellect itself originated. When it comes to natural things, our intellect must adapt itself to the thing known, because it is passive in the act of knowing, understood as the reception of species and the abstraction of universals from them. In other words: the relation between knower and thing known is non-mutual: the act of knowing does not change the thing know, while it changes the knower.

there are a few cases of attempts to integrate Scotistic themes into Thomistic philosophy, as happened in the case of Suárez.[5]

Regents are well aware of this struggle between schools within Scholastic philosophy: they often bring into the discussion Thomistic and Scotistic doctrines, and state which of them they favour. This never happens in the discussion of prime matter: we can argue that regents belong to the vast current of late Renaissance and early modern age Scotism.[6] The claim that they are consciously Scotistic is a different one: it is a fact that their theory of prime matter is grounded in Scotus's philosophy, but the thread linking the regents to Scotus is not exclusive. At the time of the regents, Scotus's doctrines on prime matter were so widely accepted that a great number of philosophers not strictly 'Scotists' successfully employed them in their philosophies: I am thinking of the afore-mentioned Suárez, but also of the commentary on the *Physics* by the College of Coimbra. Coimbrans were Thomists, yet it has been pointed out that their theory of prime matter is influenced by Scotistic solutions.[7]

3.1 Prime matter and God

Regents regarded as atheistic the theory of the identity between God and prime matter. The history of Scholasticism shows few cases of such an identity coherently claimed: one of them is David de Dinant. Clearly, what the regents reject, alongside the obvious

---

[5] I will deal with the Suárezian notion of prime matter later on. Suárez seems to agree with Scotus in many respects: for example Suárez accepts the attribution of a metaphysical act to prime matter, which is a decisive reinterpretation of the Aristotelian notion of matter as pure potency. This is required by the very notion of creation, which cannot be directed towards a being which is merely pure potency and, according to Suárez, a pure nothing, given his identification of objective potency and pure nothing. *"Quia ens in potentia obiectiva, ut ostendimus, est simpliciter nihil seu non ens actu"*, *DM*, 31, III, 6.

[6] It is now accepted by scholars that Scotism played a fundamental role in Renaissance and early modern age philosophy in general. In 2002, O. Boulnois, in his introduction to the issue of Les Études philosophiques on Scotus (*Duns Scot au XVIIe siècle*, Les études philosophiques, 2002, 1) wrote that: *"étudier «Duns Scot au XVIIe siècle» est un choix insolite et insolent. Ce numéro des Études philosophiques porte sur un objet qui n'existe pas dans les études modernes, une véritable chimère historiographique* [...] *Il s'agit de produire ici l'histoire de certaines propositions de Scot, circulant anonymement, souterrainement, et pourtant massivement, dans la philosophie du XVIIe"* (p. 1). Regarding the opposition between the philosophy of the Schools and that of the "independent philosophers": *"il n'est pas sûr que ces différents styles de vie philosophique modifient la nature des énoncés qu'ils produisent"* (pp. 1-2); and finally, that they do not want to *"faire l'histoire des perdants"* (p. 2). For a survey of Scotism across Europe, J. Schmutz, *L'héritage des Subtils. Cartographie du scotisme de l'âge classique*, *ivi*, pp. 51-81. For the same topic in a closer relation to Scotland: A. Broadie, *The Shadow of Scotus*, Edinburgh, T. & T. Clark, 1995.

[7] D. Des Chene, *An Aristotle for the Universities: Natural Philosophy in the Coimbra Commentaries*, in S. Gaukroger – J. A. Schuster – J. Sutton (eds.), *Descartes' Natural Philosophy*, London, Routledge, 2000, vol. I, ch. 2.

theological implications of this theory, is the misunderstanding at the basis of it. Thomas uses these words:

> Sic enim et oppositae differentiae ab invicem distinguuntur: non enim participant genus quasi partem suae essentiae: et ideo non est quaerendum quibus different, seipsis enim diversa sunt. Sic etiam Deus et materia prima distinguuntur, quorum unus est actus purus, aliud potentia pura, in nullo convenientiam habentes. [*CG*, I, 17, 7]

And Lamb 1613 agrees with him:

> Materia [prima] maxime recedit a Deo, quippe pura potentia ab actu puro. [TP 3]

It is clear that the identity between God and prime matter is unacceptable in Scholastic philosophy, which does not mean self-contradictory. Scholastic philosophy in its historical form is the product of many elements, the two most prominent ones being Aristotelianism and Christian revelation. But in the fifteenth century a 'Scholastic philosophy' started to gain separate dignity from 'Scholastic theology', the form of Scholastic reasoning dominant at the time of Thomas and Scotus. It is debated whether most of the Scholastics were theologians tout court or theologians and philosophers at the same time. What appears is that a Scholastic philosophy without the influence of Christian religion is hardly imaginable. What is arguable though, is that Scholastic philosophy appears to have started to detach itself from Scholastic theology, and develop on its own. The Aristotelian school of Padua may be a good example of this attitude.[8] I shall suggest that Scottish regents belong to this category of 'philosophical Scholasticism'.

What regents reject in this identity theory is precisely what most of the Scholastics reject: what is pure act (God) cannot be identical with what is pure potency (prime matter): the two beings are as far away as possible from each other in the scale of being and reality. Yet

[8] On the Padua Aristotelians, G. Piaia (ed.), *La presenza dell'aristotelismo padovano nella filosofia della prima modernità*, Rome-Padua, Antenore, 2002. On Paduan Scotism: C. B. Schmitt, *Filippo Fabri's* Philosophia Naturalis Io. Duns Scoti *and its Relation to Paduan Aristotelianism*, in *The Aristotelian Tradition and Renaissance Universities*, London, Variorum Reprints, 1984, chapter X. Schmitt claims that Fabri's attempt to create a textbook in natural philosophy *ad mentem Scoti* is important for two reasons: 1) it shows the increasing influence in Padua of metaphysics and theology from the sixteenth century on, in a curriculum which was traditionally oriented towards the arts and medical studies. 2) it reveals the importance of Scotus in the period, since Fabri sought to export Scotism in natural philosophy, an area to which Scotus did not dedicate extensive attention. Regarding the graduation theses, Schmitt's intuition of the intrinsic difficulties of a 'Scotistic natural philosophy' reflects well the fact that Scotus is the main sources for the metaphysics of the regents, but is significantly less important in natural philosophy.

the ground for this identity is implicit in the interpretation of the terms 'pure act' and 'pure potency'. Pure act is something undivided, simple, completely actuating its essence, devoid of change; pure potency could be described with the same words since it is absolutely simple, it is undivided, it is its own essence, it does not change in the sense that it is always identical with itself. In a Thomistic context the identity theory must be rejected because it is incompatible with the Thomistic doctrine that act implies form. In the wide family of Scotism (including the regents) there is the theoretical support for the theory in the claim that prime matter has its own act, before and without any form at all.[9]

A further possible support for this claim is the argument by which both God and prime matter can be reached: the *via negationis*. If the negative theology is a proper way to speak about something which is unknowable in its essence, God, then it might also be a useful tool to analyse prime matter, given its metaphysical status.

3.2 Prime matter and *actus entitativus*

All regents agree on the notion of pure potency as essential to prime matter and their enquiry focuses on whether the attribute of 'pure potency' is the whole essence of prime matter. In Scholastic philosophy there is difference between the reason (*ratio*) and the essence of something (*essentia*): the reason is what our intellect perceives as belonging necessarily to something. It is what (*quid*) we understand something to be. Essence is the metaphysical counterpart of *ratio*: it is what (*quid*) something is. Regents tend to use these two terms as synonyms, justifying this behaviour on the grounds of the identity which, they say, holds between essence and ratio: for instance, in natural philosophy, given the epistemological theory of the *species intelligibiles*, our knowledge is reliable when correctly directed towards its proper object and in this case reason and essence can be said to coincide. Of course, this does not entail that we have an exhaustive knowledge of the essences: but it does entail an accurate one.

When it comes to prime matter the problem is similar, yet made more complicated by the remoteness of prime matter from our senses. Baron 1627 claims that:

9 *"In hoc autem insania David de Dinando confunditur, qui ausus est dicere Deum esse idem quod prima materia, ex hoc quod, si non esset idem, oporteret differre ea aliquibus differentiis, et sic non essent simplicia; nam in eo quod per differentiam ab alio differt, ipsa differentia compositionem facit."* This is Thomas Aquinas' opinion on David de Dinant, in *CG*, I, 17, 6. See also D. Des Chene, *Physiologia*, Ithaca - London, Cornell University Press, pp. 94-95.

> identica est haec praedicatio, *Materia prima est pura potentia*; qualis nempe est praedicatio definitionis de definito. [TP I.1]

'Prime matter' and 'pure potency' are coextensive expressions: according to Baron, saying that prime matter is pure potency is merely the predication of the definition of the *definiendum*.

Clearly, this is just a starting point. We can draw a parallel between the definition of man and the definition of prime matter: saying that 'man is a rational animal' does not tell us anything about its actual existence, we do not go beyond the essence, which implies existence only in the case of God. The further step is to enquire into how prime matter exists, once posited that it does exist and is pure potency. Regents do not accept the Thomistic framework, which would compel them to stop the enquiry at this point: according to Thomas, prime matter is pure potency and exists only in a compound of form and matter. On the contrary, regents put in place a metaphysics of essence. The question about prime matter does not move in the direction of the Thomistic 'act of being' (*actus essendi*), but in the direction of a deeper analysis of its essence. In fact, in a metaphysics of essence there is no real distinction between an essence and its being, as Thomas claims: the essence is its existence, as Scotus says in *Ordinatio*, IV, d. 13, q. 1, n. 38. Therefore, the question of the existence of an essence must be answered by the analysis of the essence itself.

One more element is important: in the regents' Scotistic approach, we can argue that they share Scotus's theological concern about the nature of a positive object of the creative act of God, even if they never explicitly bring up this point. Scotus believes that for something to be the direct object of creation, it must be more than pure potency, it must be actual at least in a minimal sense.[10] Scotus thinks that Thomas's theory of prime matter as both created and receiving all its actuality from form leads to a contradiction.

It is interesting to see how regents employ principles proper of Thomistic philosophy (as they openly admit) to make them their own, and reinterpret them according to their philosophy. Many regents quote the Thomistic principle *potentia semper ad aliquem actum refertur*: Thomas understands this principle as evidence for the necessary information of prime matter by form. Regents on the contrary apply this principle within prime matter, looking for an act proper to prime matter within prime matter. Contrary to the Thomistic metaphysics of the *actus essendi*, the regents' metaphysics of essence leads them to prove

---

[10] "*Omne ens pendens, et productum a Deo secundum esse aliquem habet etiam actum congruum ad ipsius esse.*" Carr 1617, TP I.

the internal coherence of the notion of prime matter without appealing to anything external to it - as the act of form is.

Potentia semper ad aliquem actum refertur conjoined with *modus operandi semper sequitur modum essendi* are the two key-principles in the search for the mode of existence of the essence of prime matter. How something operates must follow from what something is. Prime matter is 1) the object of a positive act of creation; 2) the subject of information by form. In both cases, prime matter must be 'something' in order to be passive in response to an act performed on it. Passivity is one of the ten categories, so ultimately one of the ten irreducible ways beings are. 'Being passive' is a 'positive' way of being. Regents claim that there is an act proper to prime matter:

> Actus igitur materiae primae non et formalis et perfectus (habet enim a forma quod sit hoc aliquid formaliter) sed objectivus seu entitativus, per quem est id quod est extra nihil et suas causas. [King 1612, TP 2.IV][11]

This passage can be taken to represent many by other regents. Some key doctrines are being employed by King in few words. First, the tie between act and form is rejected: King speaks of an act which is not 'formal and perfect', therefore it does not come from form. In fact, regents do not deny that prime matter receives from form *quod sit hoc aliquid*, which means 'to be something determined': without form prime matter is still undetermined, essentially potential. Nonetheless, King explains that this indetermination of prime matter cannot signify the whole of its essence. Prime matter has an act proper to it, which is labelled *entitativus* or *objectivus*, by force of which prime matter *est id quod est extra nihil et suas causas*. This act (which is not the formal act) makes prime matter be non-nothing, makes prime matter be outside its own causes: these two aspects (being non-nothing and being independent of its own causes) are jointly the conditio sine qua non of even the weakest possible substance: for example, an accident does not have these characteristics, because its being is secondary and dependent on a subject.

[11] Eustachius holds the same theory: he writes that prime matter *"non tamen esse ens completum in ullo genere, quia non constat ex actu et potentia ejusdem generis, sed ex potentia Physica et actu Metaphysico seu entitativo"* (*SPhQ*, III, disp. II.I, quaestio III).

Robert Baron[12] is the author of arguably the most complete metaphysical work written in Scotland in the first half of the seventeenth century: his *Metaphysica generalis* (1658, published after his death) is an exposition of general metaphysics; it draws heavily on Suárez's *Disputationes metaphysicae*, and its main aim is completeness rather than profundity. Yet, the work is invaluable as providing a broader and more detailed view on the Scholasticism of the regents, whose *Theses philosophicae* are works written as handouts for oral *disputationes*, and not as exhaustive treatises. Baron helps us to define the concept of entitative act used by King:

> *Absolute primus* est *esse* Essentiae, et commode dici potest *Actus entitativus*; *Actus secundum quid primus* est *esse* Exsistentiae, id est, *esse* acceptum pro exsistere, et dicitur *Actus entitativus*: est autem ille *Actus Entitativus* vel rei completae et totalis, vel rei incompletae et partialis. *Actus entitativus completus* in rebus materialibus dici potest *actus formalis*, quia competit rebus materialibus ratione formae perficientis materiam et eam determinantis ad certam speciem corporis Naturalis; *Actus incompletus*, qualis est exsistentiae Materiae primae per se consideratae, non habet aliud nomen praeter generale nomen *Actus entitativi*. [sectio VII]

So, the act proper to prime matter is proper to the essence of prime matter, because that act is proper to any essence whatsoever. Lundie 1627[13] explains this important theory: *"essentia et quidditas alicujus est sufficiens intrinseca essendi ratio."* Every essence is intrinsically sufficient for existence, thanks to its internal non-contradiction, which makes it intrinsically possible. Regents do not accept the theory of the Thomistic *actus essendi*, which implies that essences participate in existence: they unanimously agree that existence

[12] Robert Baron (1596-1639) was a minister of the Church of Scotland, a theologian, a philosopher and a member of the Aberdeen Doctors. Baron graduated in 1613 at St Salvator's College, St Andrews, in the class of W. Lamb: Baron's name is listed in Lamb's *Positiones aliquot logicae*, Edinburgh 1613. He taught philosophy at St Salvator's, and graduated two classes: theses of 1617 and 1621 (the *DNB* reports Baron's departure from St Andrews in 1619, inconsistent with the 1621 theses). Baron was appointed professor of Divinity in 1625 at Marischal College, Aberdeen, where he joined the Aberdeen Doctors, and supported the religious policies of the king against the National Covenant. He died in 1639 on his way back to Scotland after an exile due to his refusal to sign the covenant. His main works are: *Philosophiae theologiae ancillans* (1621); *Disputatio de authoritate sacrae scripturae, seu, De formali objecto fidei* (1627, which originated the dispute with George Turnbull); and *Metaphysica generalis* (posthumous, 1657-8). *DNB*.

[13] John Lundie (1600-), regent. MA at King's College in 1622, under Andrew Lunan. Lundie's name appears on the title page of Lunan's *Theses philosophicae*, 1622. The *DNB* reports his appointment as regent for the year 1626, while an earlier appointment in 1625 is more convincing, since Lundie authored the 1626 graduation theses for King's College. The *OG* remarks that Lundie's class of 1627 graduated with A. Strachan: yet, the 1627 graduation theses bear the name of Lundie. Unlike most of the other Aberdeen regents, Lundie signed the National Covenant in 1638. He is recorded alive at least until 1655. *OG*, p. 54 and *DNB*.

is a mode of an essence, not external to the essence. The essence of something has all it needs to exist, provided that it does exist.[14]

Baron then adds an important qualification. The act proper to prime matter is incomplete, because the complete one *"competit rebus materialibus ratione formae perficientis materiam"*: in prime matter there is no form, consequently nothing flowing from form - that is, the perfection typical of natural things, which is due to form.

In summary, regents agree on three general points about prime matter: 1) it is actual, in the sense that it has an act; 2) this act follows from the essence of prime matter, and it is not an *actus essendi*; 3) the essence is the sufficient cause for the existence of a being - provided that the being exists. The notion of pure potency considered alone in their metaphysics of essence leads to contradictions, the main one being *"nulla potentia absque transcendentali ordine ad suum actum quidditative cognoscitur"* [Barclay 1631, TP II.2]; 5) what we know about the essence of prime matter is enough to establish what prime matter is, and that it exists.

3.3 Essence and existence

The theory that the essence of a being is the sufficient *essendi ratio* of that being is deeply rooted in the metaphysical theory of the identity of essence and existence. Usually regents prefer to express this point in a negative way: 'it is not true that essence and existence are distinct as thing from thing' (*distinctio realis*), which is the kind of distinction which grounds the principle of separability and determines whether two beings (in general, not only two *res*) are two really separate things. Regents draw this theory from both Scotus and Suárez. *Omnis determinatio est negatio*: the negation of a real distinction is the affirmation of the identity: so, if the essence and the existence of two beings are not distinct as thing from thing, then they must be identical. Regents recognise some degrees of distinction within the identity of two beings: in fact, without having two really distinct things, there are distinctions between a mode and its subject (*distinctio modalis*) and distinctions only grounded in our concepts, which are not really in things (*distinctiones rationis*).[15]

As I will argue in part I, chapter 4, about quantity and extension, the regents' interpretation of the *distinctio realis* is a central feature of their philosophy. The identity

[14] This doctrine may lead to the position that the existence of essences necessarily flows from their essence. This is not correct, because finite essences do not enjoy a perfect simplicity with respect to themselves: because they are composite, any sort of 'ontological proof' must be ruled out. Essences, with the exception of the divine essence, are possible and never necessary.

[15] Suárez, *DM*, 7, I, 4 and 18-19.

between essence and existence is a fundamental point, but what also matters is the kind of identity between them. The regents' theory of the distinction of reason between essence and existence is a qualification of such identity: I take this passage from the metaphysical section in Monroe 1632:

> Ergo essentia et existentia creaturae non differunt re sed ratione tantum, cujus fundamentum est imperfectio creaturae, quae hoc ipso quod a se non habeat esse sed ab alio participatum, intellectui humano praebet ansam praescindendi essentiam ab existentia, cum interim nec in statu potentiali, nec in statu actuali realiter distinguantur. [TM I.6]

The existence of an essence (its *esse actu*, 'being in act') is the actuality of the essence, it is not something added to it (*superadditum*) as something different that an essence *has*; consequently, an essence is not distinct from its existence as thing from thing (*distinctio realis*). The essence of a being is the key notion around which everything revolves: it appears that we are, generally speaking, in a Scotistic framework.

A similar theory is found in Eustachius, who claims that existence is a:

> modus quidam essentiae intrinsecus quo formaliter res dicitur esse actu sive extra suas causas [*SPhQ*, IV, II, II, IV].[16]

Another passage by Suárez shows the extent of the agreement between him and the regents. Suárez affirms that:

> in creaturis existentiam et essentiam distingui aut tanquam ens in actu et in potentia, aut si utraque in actu sumatur, solum distingui ratione cum aliquo fundamento in re, quae distinctio satis erit ut absolute dicamus, non esse de essentia creaturae actu existere. [*DM*, 31, V, 13]

The formal distinction with a foundation in things is enough to claim that having existence in act per se is not essential to finite beings: this avoids the metaphysical claim that existence is a necessary mode of essence, absolutely speaking. Only God has existence per se. This distinction suffices for the task because it has a *fundamentum in re*; which means that our concepts tell us about a distinction which is not really in nature (real

[16] Already quoted above, section 2.

distinction) but which is still grounded on how things are. Thus, the essence is not its existence, yet it is not distinct from it in reality.

When it comes to prime matter the same principle holds: the essence of prime matter is not its existence (that means, prime matter is not necessary), nonetheless, the existence of prime matter is distinct from its essence only by reason - the weakest degree of distinction.

4. Prime matter as receptive entitative act

According to the regents, the analysis of the essence of prime matter is not complete without a further qualification: prime matter is a *receptive* entitative act.

Regents agree that: 1) prime matter exists (quod sit); 2) prime matter is actual in a metaphysical, non physical, sense of the term: it has a metaphysical, not a formal act; 3) prime matter's existence is a mode of its essence. The essence of prime matter now requires a qualification which enables the regents to claim more precisely what it is (quid sit). In fact, saying that something is 'actual' does not convey any information about what this act is an act of. Being actual is a formal aspect of prime matter, still in need of a material aspect.

The answer is once again agreed upon by the great majority of the regents, despite some variations in terminology. Prime matter is pure potency, actual in virtue of its essence of being pure potency; finally this pure potency is spelled out as 'receptive'. The metaphysical role of prime matter is to receive forms, to be the subject of the information by forms. The concept of pure potency alone is not considered sufficient to express this essential openness of prime matter towards the form. The qualification of 'receptive' highlights again the difference with the Thomistic solution. In fact, according to Thomas the definition of prime matter as pure potency is sufficient to claim such openness towards the form: prime matter is devoid of all acts (which always come from form), therefore a 'pure potency' is always directed towards a form. On the contrary, the regents attribute an act to prime matter: thus, they have to introduce a further qualification to explain why prime matter is directed towards form.

The opposition by the regents to the Thomistic slogan that an act is always related to a form becomes even more explicable when regents seek to expound the sort of pure potency that is proper to prime matter. The focus moves from the essence understood as an act (which grounded the claim that prime matter enjoys independent existence) to the essence understood as a potency (which introduces a better definition of this potency). There is clearly no contradiction between being in act and in potency for prime matter, because prime matter is in act with respect to a metaphysical act, while it is in potency with respect to a formal (physical) act: there is contradiction only if the same subject is in act and in potency at the same time in the same respect. A Thomist would probably argue, on the basis of his own metaphysics, that in this case contradiction is avoided only in words, not de facto: prime matter, when informed in a compound, would obtain a physical act from form which would be added to the already existing metaphysical act. Then, the result would be a compound that is not essentially one, since from two beings in act no unity per se can be obtained, but only a unity per accidens. The Thomist's objection only holds if we accept the theory that all acts come from form. Regents reply to this objection by accepting the idea that:

> ex duobus actibus imperfectis potest unus perfectus consurgere [King 1612, TP 2.III].

Where does the strength of this reply lie? Regents accept the distinction between perfect and imperfect act (or, to put it differently, complete and incomplete act). A complete act belongs to a complete substance, which is only the result of the union of form and matter. An incomplete act belongs to the components of the compound, which are in need of each other in order to yield a complete substance, but which are actual per se for the reason that only something actual (at least actual in a metaphysical sense) can get into composition with something else. If we accept the notion of incomplete act, then the claim that a component of a substance is both in act and in potency at the same time is not contradictory, because it is not in act and in potency in the same respect.

A corollary of this theory is that:

> non omnis potentia subjectum praesupponit. [...] aliquam potentiam substantiam esse et non qualitatem [Reid 1614, TP 1, III-V].

This principle moves the analysis one step further. I am not sure how to read 'quality' in this context: I shall suggest two interpretations. 1) Reid is perhaps not taking the term 'quality' in its categorial sense, but rather in a more general sense of 'attribute'. In fact potency can hardly be reduced to the category of quality, which would restrict the predication of potency only to accidents falling under the notion of quality. 2) Perhaps Reid is here employing a Suárezian terminology. Suárez distinguishes between transcendental potency and predicamental potency: the former is the objective potency, proper to a possible thing, which we have seen before; the latter belongs to the second species of quality, and is the real potency, either active or passive. This potency refers to quality because only qualities are proximate principles of the actions in creatures.[17]

In both cases, the separation from the Thomistic theory is here complete: potency as such does not necessarily require a subject because potency can be a subject, as it is in the case of prime matter. This passage is not in contradiction with the principle that every potency must be referred to an act, because Reid holds that prime matter has an act on its own, and is also pure potency.

A fundamental metaphysical distinction without which all the previous theories are in some sense groundless is to be found in many theses: the distinction between 'pure potency' and 'in pure potency' and conversely between 'act' and 'in act'. Wemys 1612 writes that:

> distinguendum est itaque inter actum et esse actu, et potentiam et esse potentia. [...] dubium idcirco est an satis philosophice dici possit vel materiam esse potentia vel formam esse actu. [TP 6, III-IV]

The second part of the passage is meant to be an explicit attack against Thomistic philosophy. The philosophical relevance of this theory is that the couples: form/act and matter/potency are finally overtaken in a metaphysics of essence. There cannot be an univocal sense of act and potency; act and potency are not coextensive with form and matter; according to the regent, it might be even possible to say that form is not act (because it is essentially open to potency) and that matter is not potency (because its potency relies on prime matter being a metaphysical act). It is then more accurate to say that prime matter is *"pura potentia, non in pura potentia, in actu et non actus"* (Wemys 1631, TP I.2). This is a distinction common to late Scholasticism: it was formulated to make sense of the specific metaphysical status of prime matter.

One final qualification helps the regents to finally give a complete definition of the essence of prime matter: it is based on the distinction between *potentia objectiva* and *subjectiva*. This is how Baron addresses the point in his *Metaphysica generalis*:

> Sub hac *Potentia Logica* continetur *Potentia* illa quam *objectivam* vocant; ea enim a parte illius rei quae dicitur esse *in potentia objectiva* respectu causae, nihil aliud significat quam non-repugnantiam ad produci a tali causa, i. e. significat non impossibile esse ut illa res a tali causa producatur: unde patet

[17] *DM*, 42, III, 10, quoted in Leopoldo Prieto López, *Suárez, crocevia nella filosofia tra medioevo e modernità*, Alpha Omega, IX, 2006, No. 1, pp. 3-38, pp. 29-30.

> *Potentiam* hanc *objectivam* non esse realem, tum quia consistit in negatione impossibilitatis (negatio autem non est Ens reale, sed formaliter non-Ens) tum quia haec *Potentia objectiva* competit rebus antequam a Deo ipso producantur, nihil autem, absolute loquendo, *reale* rebus competit antequam a Deo accipiant esse. [sectio VII]

Wemys 1631 integrates this passage by saying that prime matter is not an objective pure potency because this would prevent it being an entitative act. Thus, prime matter is *subjective* pure potency. This notion is indebted to the philosophy of Scotus. A subjective potency is a potency which already exists, while an objective potency, as Baron explained, is merely a logical possibility.[18] This reminds us of the Suárezian distinction between transcendental and predicamental potency. Prime matter cannot be objective potency because it has a metaphysical act, has its own existence, and is not a mere logical possibility: rather it is the root of any predicamental possibility. Or, in other words, it is a metaphysical possibility.

The essence of prime matter can now be stated in its full form: prime matter is *subjective receptive pure potency*. This definition is the product of the different positions we find in all the *Theses philosophicae*. Its philosophical content is shared by all regents, while its form is subject to some variations. As I said, I think that these versions differ in form, not in content:

> Ratio principis materialis est potentia universalis recipiendi omnes formas indistincte. [Adamson 1600, TP V]
>
> Materia prima essentialiter est substantia incompleta, et pura potentia subjectiva (cui tamen actus entitativus competit). [Forbes 1623, TP II]
>
> [Materia prima] est pura potentia receptiva, non potentia objectiva, cui opponitur actus metaphysicus. [Barclay 1631, TP I.3]
>
> [materia prima est] pura potentia passiva. [Dalrymple 1646, TP IIII]

[18] By 'logical possibility' Baron here means, alongside Scotus and Suárez, the sort of possibility which an essence has before existing, before being created. According to Suárez, this possible essence is still a pure nothing. After being created, an essence is a subjective possibility, or, which is the same, a real possibility.

One very interesting formulation is given by Fairley 1623, TP I.3: the causality of prime matter is *'passiva actuatio potentiae.'* I suppose that this last quote clearly shows the level of sophistication reached by the regents in their metaphysical theories, and the difficulties that students had to master.

5. Conclusion

In this chapter I have dealt with the definition of the essence of prime matter in the *Theses philosophicae*. All regents agree on the definition of the essence of prime matter as 'receptive entitative act'. 'Receptive' because prime matter is essentially open to form. 'Entitative act' because prime matter has a metaphysical act proper to it, an act which is prior to and independent of the act of form. Prime matter is pure potency, which means that its essence is being pure potency. Yet, the regents are influenced by Scotism, and go beyond the Thomistic definition. Just as something that is 1) the direct object of an act of creation, and 2) a component of a substance, must be actual, in the same way prime matter must necessarily be actual, because a pure potency devoid of any actuality is only a logical, not metaphysical, possibility. I believe that the regents are also influenced by Suárez, as is evident from terminological and doctrinal similarities. Suárez himself was influenced by Scotism, so that it can be argued that Scotism exerted influence on both the regents and Suárez.

The analysis of the essence of prime matter, in particular with regard to the relation between essence and existence, has also shown that the regents hold a metaphysics of essence, once again departing from Thomism.

De potentiis materiae primae

In his *Summa philosophiae quadripartita* Eustachius a Sancto Paulo structures the analysis of prime matter around three philosophically distinct aspects which together give us a complete account: 1) that prime matter is, and what it is (quod sit and quid sit); 2) what the potencies (*potentiae*) of prime matter are; and 3) what the properties (*proprietates*) are. I intend to follow the same scheme for two reasons. First, it is clear and consistent. Secondly, this scheme mirrors the metaphysical structure proper to prime matter, which entails, in this order, the definition of its essence as pure potency, then the explanation of the notion of potency, with the introduction of the relationship between matter and form, then the analysis of the specific contribution of prime matter in the compound. Thus, the *ordo expositionis* follows the *ordo essendi*, the metaphysical order of the thing expounded.

These are Eustachius's words about prime matter's potencies. Eustachius writes that:

> Materiam primam secundum se spectatam aiunt omnes omnium formarum expertem esse ac simul omnium capacem esse; sive, materiam esse in potentia ad omnes formas: Quare materiam ipsam appellant *potentiam*; est enim hoc essentiale materiae. Consistit autem in duobus passiva ista materiae potentia: primo quidem in eo, quod ex materia possunt formae materiales virtute naturalium agentium educi; secundo in eo, quod illae omnes et nonnullae aliae quae ex ipsa non educuntur, possunt in eandem recipi: Sicque potentia materiae partim Eductiva partim Receptiva dicitur. [*SPhQ*, pars III, tractatus I.II, quaestio III]

As we have seen in the previous chapter, the *Theses philosophicae* agree on the definition of prime matter as pure potency. Further qualifications are that prime matter is a receptive entitative act and that it is in a state of subjective possibility towards existing, since it is a metaphysical act, not just a logical (merely non-contradictory) being. Regents engage in the exposition of what follows from this still general analysis, namely: 1) in what way prime matter is indistinctly open towards any form; 2) what is the appetite (*appetitus*) of prime matter for form in general, and 3) what it means that forms are educed (*eductae*) from prime matter. This analysis is the intermediate moment between the analysis of the notion of prime matter as pure potency and the analysis of the proper contribution of prime matter in the compound (quantity, extension, incorruptibility, which is the subject of the next chapter). In a way, it is possible to say that in the first two chapters prime matter is regarded as passive towards form, while in the third it is regarded as "active" towards form. Inevitably, some aspects will be fully meaningful only at the end of the analysis, which is also one of the premises for the account of Transubstantiation, the subject of chapter 4.

In this chapter, I follow Eustachius's division of prime matter into *'partim receptiva'* and *'partim eductiva'*, 'in part receptive' and 'in part eductive'. The powers of prime matter show its role of material subject of all forms, which is receptive when receiving forms but also eductive when forms are drawn from it. Only material forms are drawn from prime matter: the first problem that I shall investigate is the relationship between the rational soul and prime matter. The receptiveness of prime matter is then analysed in terms of appetite towards form. This raises the question whether the appetite of informed prime matter can be said to completely satisfy the potency of prime matter, which is essentially open towards form. The second part of the chapter is dedicated to the concept of the eduction of forms. The last part deals with the *Theses philosophicae* by Dalrymple, written in 1646 for the students of the University of Glasgow: the regent puts forward an interpretation of eduction which I believe is influenced by the early modern philosophy of the period.

1. *Partim Receptiva*: prime matter and form

In Scholastic natural philosophy, prime matter cannot exist alone without form. We have seen in the previous chapter that some sort of existence must be attributed to prime matter: precisely the sort of existence of the material component of a physical substance. Prime matter is not a substance in the way that a physical substance is, because it is not a complete substance. Yet, in order to enter into composition with form, prime matter must have the incomplete existence proper to an entitative act, whose essence is pure potency. This means that form is the natural completion of prime matter, and conversely, that form can be truly form only when informing matter. This is a general Aristotelian principle, which regents do not reject.

1.1 The problem of the rational soul

In the context of a Christian reading of Aristotelian philosophy one problem is immediately evident: the status of the rational soul (*anima rationalis*) with respect to existence independent from matter and with respect to its origin within the compound. The problem originates from the double relationship that the rational soul has towards matter as form-in-matter (the rational soul is the substantial form of a physical compound) and towards independent existence, as the human soul is said to be immortal in the Christian tradition.[1] In fact, rational souls (or, the substantial forms of men) must survive the destruction of their physical compounds, if they are to resurrect from death and reincarnate. This is the doctrine of the resurrection of the bodies.

The regents do not reject this view, which is shared by all Christians. This belief raised fundamental philosophical questions. The debate in Scholasticism included the interpretation of the most significant passages in *De Anima* III by Aristotle: the Stagyrite writes that some activities of the rational souls are independent from matter, in the sense that they reveal operations which do not depend on matter. They perform, for example, the knowing of the universals. In Scholastic epistemology, the knowledge of universals is obtained through the process of abstraction of the essence of things from their individual material being, an operation by the agent intellect which acts on the material offered by the possible intellect, which receives the notions from the common sense (*sensus communis*), a sort of unified sensorial perception posterior to the five senses. If only compounds are individual, and if only individuals really exist,[2] then the process of abstraction goes beyond materiality and must be a sign of an immaterial principle of activity: the rational soul. Prime matter, as we have seen, is the root of potency, and potency implies corruptibility, because for a potency to be realised the former act must be corrupted. Therefore, immateriality goes with incorruptibility. Something incorruptible is immortal, and ultimately simple.[3]

[1] The theory that the rational soul is the substantial form of man received an official endorsement by the Roman Church during the ecumenical Council of Vienne, 1311-1312, chaired by Clement V. This council, famous for the condemnation of the Templar order, is also crucial for the acceptance of a philosophical and theological theory in the teaching of the Roman Church. One of the decrees states that the rational or intellectual soul is the form of the human body of itself and essentially. Whoever rejects this theory is to be considered a heretic.

[2] King 1620, TP I: *"nulla datur entitas in communi, nisi determinata per entitatem particularem alicujus speciei; nec potentia in communi nisi determinata per particularem potentiam."*

[3] This is the case of the celestial bodies, part II, chapter 3.

Whatever the original position of Aristotle on this point was (there is in fact a vast debate on whether Aristotle agreed with the immortality of the individual soul or not),[4] Scholastics interpreted his words as adaptable to the Christian dogma of the immortality of the soul. Regents show a twofold approach: on the one side, they hold that the human soul is immortal; on the other, they usually include the analysis of *De Anima* III in natural philosophy. They regard the analysis of substantial forms as belonging to natural philosophy, and this includes human souls. This Aristotelian approach is generally stronger than the distinction between substantial forms in material and immaterial based on their activities. These distinctions are not enough to dedicate a branch of philosophy to the exclusive exposition of the characteristics of the rational soul.[5]

It may be noted that this analysis of the human soul as the substantial form of men does not include what today we call 'theory of knowledge', which is a part of logic in the theses. It appears that in Scholastic philosophy the immortality of the soul is justified in virtue of the investigation of our knowledge of the universals. In the natural philosophy of the theses, the regents do not attempt to prove the immortality of the soul, which is already proven through the revealed word. Rather, given the immortality of the soul, the regents seek to analyse the human soul according to each specific branch of philosophy: when its activity is knowledge, it pertains to logic, when it is the information of matter, it pertains to natural philosophy. I believe that this approach is a sign of a deep conviction by the regents that, in general, matters of faith are rarely if ever proven in philosophy. Thus, the separation of spheres between theology and philosophy is clear.[6]

This could be regarded as a further claim for the independence of natural philosophy within its own sphere: in the theses, different areas of philosophy rarely overlap.[7] Necessarily, prime matter implies notions which are also dealt with in either logic or metaphysics, but this is due to the specific nature of prime matter, a component of physical compounds which is not given to us as a direct object of knowledge. The regents' deployment, in natural philosophy, of rational souls as forms-in-matter allows for a general inclusive account of the relationship between forms and matter, without the need of making a distinction where regents did not want to.

---

[4] I will deal with the regents' reception of Aristotle on this point in the Conclusions, section 2.

[5] In the second half of the century, under the influence of Descartes, regents developed some themes that we find in St Andrews graduation theses in the first half of the century, regarding a metaphysics of separate substances. This branch of metaphysics is called pneumatology.

[6] This theory is put to a test in two following chapters, first about Transubstantiation (part I, chapter 4), then about the role of natural theology in the theses (part II, chapter 3). In this chapter, I shall return to this point when dealing with the eduction of forms and the role of God, section 2.2.

[7] I argued for a similar claim with respect to the importance of the argument from natural philosophy in the demonstration of the existence of prime matter, in chapter 1, section 1.2.

1.2 Prime matter as openness towards form

'Being open to form', or 'being open to information', means that prime matter is a principle of the physical body. Not simply per accidens but essentially, because it is a necessary component of the compound. The other essential principle is form. The traditional third principle is privation (*privatio*): privation, according to the regents, is the absence of form. In a body in becoming, the actual form entails the presence of the absence of another form, and it also entails that no specific form can essentially belong to a compound: if it were so, a compound would be a necessary being. Regents usually consider privation as a third principle per accidens of becoming, in agreement with Aristotle. They also tend to include the analysis of privation in the analysis of form, because privation can also be regarded as the presence of the previous form. Mercer 1630 stresses this last point:

> Privatio non tam est absentia formae subsequentis, quam praesentia formae praecedentis, non quidem qua forma est, sed qua materiam praeparat ad formam subsequentem accipiendam. [TP IV.2]

The logical and metaphysical status of prime matter is thus fully understood only in the context of natural philosophy. This is why regents never deal with prime matter in logical or metaphysical sections, even if these are fundamental as introductions to key notions of natural philosophy.

Prime matter is the passive principle, while form is the active one: despite the attribution to matter of an act and attributes, regents do not go beyond the Aristotelian viewpoint that all composite beings are the result of an active principle acting on a passive subject. From this point of view, it is interesting how close the definitions of matter and substance can be understood to be in their different levels. For instance, in *Met.*, V, 8 1017 b 23-26 we read that substance is a) *"that which is the ultimate underlying stuff, that is not predicated of anything else"*, and b) *"that which, being something determined, can also be separable, and this is the structure and form of any thing."*[8] When making a parallel between this definition and the attributes of prime matter in *Phys.* I, the notion of 'subject' of predications is shared. In the first case, substance is intended in two senses, one logical

[8] a) '[T]ό θ' ὑποκείμενον ἔσχατον, ὃ μηκέτι κατ' ἄλλου λέγεται'; b) 'τόδε τι ὂν καὶ χωριστὸν ᾖ· τοιοῦτον δὲ ἑκάστου ἡ μορφὴ καὶ τὸ εἶδος.' My translation.

(subject), and the other one logical and metaphysical (something determined and separable [from something else]). Prime matter per se is only a metaphysical principle, because it does not belong to the category of substance: it cannot exist 'separate' from anything else. We will see that regents get away from this theory when claiming that prime matter too is a 'substance'.

Two essential principles of the compound yield a unity per se. The tie between matter and form is as strong as an essential unity, yet it is in no way a relation of identity.[9] This unity entails that form and matter cannot exist one without the other:

> Nulla forma physica habet modum essendi independentem a materia nisi anima rationalis. [Fairley 1623, TP III.2]
>
> Materiae essentiale est et necessarium formam semper appetere. [Wemys 1612, 5.III]

These two passages are representative of the viewpoint of the *Theses*. In Fairley the terminology is proper to a metaphysics of essence ('modum essendi'): it is not incompatible with what has been said about prime matter having existence per se in virtue of its metaphysical act. In fact, Fairley is referring to the physical world, where a metaphysical act is not enough to sustain existence. So, it is only the compound which really exists, even if prime matter has a mode of existence, which is not the mode of existence of a complete substance. In different words:

> Cum diversae numero formae non possint eandem numero existentiam tribuere, non omnis existentia materiae est a forma sed completa tantum. [Stevenson 1625, TP XII.4]

We find again the principle that form and matter are two distinct metaphysical entities, both actual, yet incomplete. Aedie 1616, TP I, expresses the role of matter with respect to becoming by saying that matter is the principle of being and non-being of all perishable things, and showing that forms cannot exist unless in matter but that prime matter is the root of potency. The metaphysics of act and potency of the theses seems to hold onto the traditional view of degrees of being, where the spectrum extends from God to prime matter. God as supreme being is pure act, creatures are a composition of act and potency,

[9] Rankine 1627, TP VI.5: *"materia enim per formam determinata, et forma quatenus materiam determinat, non duo constituunt principia, sed unum."*

and prime matter is pure potency and the lowest degree of actuality. In this theory, the presence of potency is synonym with imperfection, according to the principle that:

> quae minus participant potentiae verius et magis proprie esse existimandum non est. [Wemys 1612, TP 7.III]

There are two interpretations of this principle, both accepted by regents: 1) in epistemology, it is true that the less something is in potency, the more we can get to know it ['verius esse']; 2) in metaphysics, the less the potency, the more the act ['magis esse']. As essentially pure potency, prime matter is the least known thing within the physical realm. Creatures on the contrary are open to our knowledge because of the balance between act (what they are) and potency (what they can become). The pure act is in itself the most knowable thing and the most 'real' thing, but as in the case of pure potency it extends beyond our limited comprehension, and it can only be object of a mediated and ultimately insufficient knowledge.[10]

We can also better understand the role of the *via analogica*: prime matter is known not per se, but analogically with respect to finite beings: in absence of form, the analogy holds between the act of prime matter as "form" of the compound, and the pure potency of prime matter as "matter" of the compound.

There is one respect in which matter is more perfect than form, a respect which illuminates the fundamental reason why forms cannot exist outside matter. According to Fairley 1623, forms are:

> perfectiores materiae secundum Entitatem et Essentiam, sed imperfectiores secundum essendi modum. [TP III.5]

A compound is the result of the individual contributions of form and matter: forms contribute essence, to the extent that forms can even be called the 'end' of matter; yet, matter plays the fundamental role of sustaining forms, thus it is prior to them under the concept of the 'mode of existing' [modus essendi]. It is true that matter is ordered towards form as much as potency is towards act, *"tanquam ad finem"* (*ibidem*), but it is also true that *"necessitas ad causalitatem materiae* [est] *ejus existentia"* (Wemys 1631, TP II.4). In

[10] This relates closely to the problem of what sort of philosophical knowledge of God we can have. A wide debate took place in Scholasticism, while regents rarely take an explicit stand on this matter. We can infer their position from the broader context of their epistemology, as in this case, and in the analysis of book VIII of *Physics*, where Aristotle reaches up for the immobile motor on the basis of the analysis of physical movement. Regents object that this inference is 'ill-based', part II, chapter 3, section 4.

Scholastic natural philosophy, finality is a form of causality: it is evident that, for everything to be causally active, it must exist; on this basis prime matter is the existing potency directed towards its end, form. With respect to the openness of prime matter towards forms, forms are final causes. In this context, regents claim nothing more than a constitutive openness of matter towards form. Indeed, final causality is not rejected in the theses also on a purely physical level, as is clear from the discussion of movement. In the analysis of the structure of physical compounds, form, intended as the final cause of matter, only means that form makes matter perfect in its essence, which is otherwise incomplete.

As well as matter's priority over form in respect of the mode of existing, matter is also responsible for the endless becoming that we experience: this is another contribution to the structure of compounds. It is important to remember that so far matter is only regarded as a 'passive principle', and in no way as a positive subject: we cannot say that matter positively acts on form, or that matter acts at all. The fact that matter can be said to be active is the consequence of the union with form, a union within which matter has characteristics it would not have were it able to exist alone.[11]

In natural philosophy, following the act/potency theory, form is act and matter is potency: Scholastics interpret the natural becoming as the formal active principle affecting the material passive principle, but also being affected in return by the same material passive principle. The complete substance resulting from form and matter thus is the union of two substances incomplete though in different respects, the former attributing actuality, the latter attributing potentiality, both to the compound and to one another. As Fairley 1615 claims, matter makes form *patibilis*. Form is not *patibilis* per se, but it is as form-in-matter [TP XX.2]. This term can be intended in two ways: 1) patibilis as 'sensible', belonging to the physical world; here matter is the principle of materiality and form is a material form; 2) patibilis as 'responsive to passivity', something that form, considered alone, is not.[12] Here matter is the principle of passivity, which comes from materiality. The two meanings are thus connected: the result is that form in a compound undergoes changes because of matter. What sort of changes? Principally, and this is the key notion of becoming, the bond between one form and its matter is not necessary: different forms can inform the same matter in time.

[11] I shall argue in the next two chapters that the regents make an interesting claim for a positive predication of attributes to matter; that means that matter cannot be interpreted as receiving all attributes by form in the compound, but also as having attributes on its own, namely quantity, extension and divisibility. This theory stems from the Scotistic notion of prime matter that regents deploy.

[12] Wemys 1612, TP 4: *"forma omnis incorporea est et per se indivisibilis."*

From the point of view of the formal relationship between form and matter then, matter is indistinctly open towards form: form is needed by matter as an end, without which matter is not complete. The qualification of 'indistinctly open' underlines that any form can be an end of matter, there is no a priori reason for which 'form' generally taken cannot inform any portion of matter.[13]

1.3 Prime matter's potency as *appetitus*

The notion of prime matter as a metaphysical act whose essence is receptive pure potency explains the genus and the difference (*differentia*) of it, the two terms which convey the definition. In this case, 'metaphysical act' is the genus and 'pure potency' is the difference: in virtue of these two qualifications the act of prime matter can be distinguished from any other act. The definition is in fact the predication of the essential attributes which locates something within its genus and differentiates it from other members of that genus. This is why both elements are necessary: the genus to identify the sort of being we want to define, the difference to predicate something proper to it and to nothing else in the same genus. Yet, as much as the difference 'rational' in the genus 'animal' is not enough to explain what man is, 'receptive pure potency' is not enough to explain prime matter. The notion of 'appetite' (*appetitus*) is the logically first characteristic of prime matter which is not dealt with in the definition.

1.3.1 *Appetitus* and *bonum*

Appetite is a key notion in Scholasticism. It is the second qualification of what we have so far termed 'openness' of matter towards form: the notions of receptiveness and appetite complete the analysis of the openness of matter towards form. This relationship between form and matter is just an individual occurrence of the universal appetite, which is a driving principle shared by all created beings. Thomas and the Scholastics hold that:

[13] This is a principle of general natural philosophy, the branch of natural philosophy studying the principles of natural bodies in general I am concerned with in part I. In special natural philosophy, which studies natural bodies as having differences natures (*naturae*), only specific forms can respectively inhere in the two kinds of matter: sublunar and celestial. Scholastics usually accept the theory of the different nature of sublunar world and heavens (part II, chapter 3).

> Appetere est commune animatis et inanimatis. [*ST*, I, 80, I]

Anything created strives for (*appetit*) something else, no reality is static because *"appetitus inclinationem ad bonum notat"* (Stevenson 1629, TP VI). This is a debated claim. It entails three points: 1) the good (*bonum*) is what created beings seek, because the good is what makes them more perfect; 2) thus, the good is an end for created beings; 3) 'inclination' (*inclinatio*) here means that leaning towards a good is not a fixed path from one determined starting point A to a determined final point B. Creatures, and rational ones above all, are open to different ends, which are all good formally (so the unity of the principle is preserved) but are different materially. Famously, Aristotle claimed that the nature of good depends on the substance, not vice versa. Christian theology inevitably translated the words of Aristotle in a different context, but the original idea of good is not superseded. Thus, God is the absolute good, equally good for any substance.

In natural philosophy, all beings move towards their good, formally one, materially different as the beings are different, and this is a fundamental internal principle of movement that they have, and by the acquisition of which they are completed. Any being is good, because *bonum* is a transcendental attribute of beings, along with *verum* and *unum*. Without goodness, unity and truth the very concept of being becomes empty. Form is the 'good' of matter, consequently matter strives for its end; and conversely, form is an end and matter while attaining its end at the same time attains its good. Good and end are not separable.

There is an essential directedness within all composite beings, which is due to the metaphysical structure of the compound of form and matter. Form alone would be unable to attain anything else from itself, being a good in itself (not 'the' good, of course). The union between form and matter is essentially 'one' because the composite is such per se, not per accidens, yet the union is not essential, because by constitution the potency of matter cannot be made completely actual by any determined form. This point in its relationship with appetite is addressed by Wedderburn 1608:

> formae accessu suo appetitum explere potest, potentiam non potest. [TP II.4]

Matter is essentially potency, no form can change the essence of matter by simple information of it; what form does is to satisfy in each case the appetite that matter has for form, which is the constant power of the essence of matter as potency. Or, in other words, it is the physical way in which the metaphysical potency is individuated in a compound.[14]

Regents believed they had explained the metaphysical structure of natural becoming by these key notions: potency, appetite, form, matter and privation. It appears that the role of matter is both active and passive: passive in the specific sphere of natural philosophy (the appetite being the receptiveness of matter towards form), but active on the metaphysical level, because matter's essential constitution as pure potency can never be restrained by form, and always seeks to replace the present form. Matter is truly the underlying active principle of physical becoming.

There is a crucial objection made against the theory that prime matter is indifferent towards forms: the experienced directedness of natural phenomena. Creatures belonging to the same genus tend to behave in the same way under the same circumstances; if today we account for this evidence on the basis of the principle of uniformity of nature and the concept of physical laws, regents did not have anything resembling the latter. Physical directedness is thus seen as a consequence of the principle of uniformity of nature ('nature does not move by leaps') and of the constancy of natural essences, implying that things cannot do anything contradicting their essence. The objection based on the fact of natural directedness can be fully rejected only after the explanation of natural movement, because of the role played by form as nature of bodies (part II). From the point of view of prime matter, the objection is partially answered by the distinction between matter *simpliciter spectata* and determinate matter:

> Materia simpliciter spectata non ad unam magis formam quam ad aliam propensa est, neque unquam aliam formam appetit, quia praesentem fastidit. Quare cum determinatur ad certam formam, ad eam solum habet potentiam. [Strang 1611, TP III]

Unfortunately this is the only passage clarifying the question from the point of view of matter alone. Strang seems to hold that after being determined by a form (which means, after being made the matter of a determined compound), until the compound exists this portion of matter cannot accept any other form to replace the present one. Form here means 'substantial form', the sort of form giving essence and unity to a compound, not accidental form, which can always vary without causing the compound to dissolve. So, in matter is to be found the root of the constancy of becoming within the same compound; but also the ultimate root of one compound becoming another one.

---

[14] I intend to leave aside the discussion of celestial bodies, which I deal with in part II, chapter 3.

1.3.2 Different theories on the nature of *appetitus*

Regents offer different accounts of the nature of prime matter's appetite. Appetite is analysed both in its relation to the potency of prime matter (an internal relation) and in its directedness towards form (an external relation):

> Materiae appetitus nihil aliud est, quam inclinatio, eaque passiva ad formam suscipiendam, eumque a privatione habet. Ex quo sequitur materiae appetitum re non differre a potentia. Et hinc quicquid explet appetitum potentiam perficere et contra. [Bruce 1614, TP IIII]

> Materiae appetitus est affectus habendi formam, ad quam propensione quadam suam inclinat. Appetitus igitur materiae potentiam non adaequat. [Wedderburn 1608, TP II]

These two passages differ on the issue of the relation between appetite and potency: in Bruce 1614 we read that the appetite perfects the potency of matter, because there is no real distinction between appetite and potency. In Wedderburn 1608 instead, the appetite is said not to match, satisfy (*adaequat*)[15] the very potency of matter. From which we might wonder about the distinction between appetite and potency. If the real distinction were the only logically possible distinction between entities the two passages could be mutually contradictory. Wedderburn does not make his claim on this issue, but we can complete it thanks to further qualifications of the notion of distinction.

Suárez holds that modal distinction between existence and essence is sufficient to ground the claim that the existence does not belong to the essence of something, because a modal distinction is not dependent on our intellect, but reflects a distinction in nature. So, it occupies the middle ground between the real distinction, between a thing and another thing, and the distinction of reason, which is between beings distinct only because an

[15] *Adaequatio* is a Scholastic term indicating equality, in terms of quantity or in terms of proportion. When equality is perfect, it does not admit degrees ('more' or 'less'). In the debate on prime matter and its potencies, I translate this concept with 'to match' and 'to satisfy' because the the Latin adaequatio reminds us of both meanings. The question is whether the appetite is equal to prime matter (and vice versa), and whether this appetite satisfies the potency of prime matter: that is, whether it is equal to this very potency. Adaequatio is famously deployed in the definition of truth as *adaequatio rei et intellectus* (for example: Thomas, *CG*, 1, 59, n. 2): I believe that the relation which occurs between prime matter and one of its potencies cannot be explained with the traditional translation of 'correspondence', because it is not the same relation as between the known thing and the intellect.

intellect perceives them to be so, and ceasing to be distinct were an intellect not thinking this distinction.[16] This does not mean that the intellect in question creates the distinction, or makes it real in things by thinking it. In fact, the distinction of reason can be twofold: with a foundation in things (*distinctio rationis ratiocinatae*), when the intellect is reflecting some sort of distinction between components within the same thing, or without a foundation in things (*distinctio rationis ratiocinantis*), when the act of the intellect establishes such a distinction.[17] The general realist approach of Scholasticism does not allow us to say that the object known is in any way affected by the act of the knower, because the relation of knowing is non-mutual, that is, directed from the knower to the known in a way that leaves the thing unchanged.

Thus, Wedderburn would just need a modal distinction to maintain the distinction between appetite and potency and claim that appetite does not match potency.

The two positions then underline different aspects of the same question: Bruce holds that appetite and potency are not really different and that appetite perfects potency; Wedderburn that appetite does not match potency and, if our argument is right, that they are not really distinct, because the modal distinction could suffice. And the modal distinction is not a real distinction. So, within the same theory of the non-real distinction between appetite and potency, two theories are possible, that 1) the appetite perfects potency and 2) that the appetite does not match with potency.

What about the terms 'perficere' and 'adaequare'? Regents are just expressing the same concept with different words: within the same identity, something perfecting something else is completely realised, and therefore adequate with the thing perfected. This would not follow in the case of the real distinction, for example, with a cause 'a' perfecting 'b': here the cause would not be a being adequate to 'b', but just the adequate cause of 'b'. So, appetite entails both the perfection of the potency of matter, because the potency is actualised, and the *'non adaequatio'* with potency, because the appetite does not match the whole of the potency of prime matter. Prime matter retains its potency towards another form.

Appetite is treated by regents as not really distinct from potency, because they both flow from the essence of prime matter, and a subject is not really distinct from its attributes. Alongside a relation to potency, appetite is also understood in relation to form. I called this relation 'external' because form is a principle external to the matter which is informed.

---

[16] Suárez also states the identity between the modal distinction and the formal distinction. Gilson argues that Descartes might have been influenced by Suárez in his theory of distinction (*Index scolastico-cartésien*, New York, Burt Franklin, 1964, text 148).

[17] Suárez, *DM*, 7, I, 4.

When it comes to the qualification of this appetite in relation to form, regents vary their responses. Stevenson 1629 offers an interesting explanation of the concepts at work here:

> Desiderium est de bono absenti, complacentia de bono praesenti, privatio carentiam boni, appetitus denique inclinationem ad bonum notat. Ergo appetitus abstrahit a bono praesenti vel absenti. Adeoque a desiderio, complacentia et privatione. Appetitus est universalior privatione, et prior secundum rationem. Privatio, cum sequatur appetitum non potest esse eius causa. Et cum praecedat desiderium, medium locum tenet inter appetitum, et desiderium. [...] materia per naturam appetit bonum, divinum, et appetibile. Ergo primo et per se appetit formam, per accidens etiam privationem ei junctam. [TP VI]

The context indicates the influence of moral philosophy in terminology. Modern philosophers who opposed the schools invariably pointed out how this "overlapping" of disciplines, due to the role in natural philosophy of concepts such as good, final causality, end and appetite, was an unacceptable anthropomorphic tendency. This terminology was abandoned outside Scholastic philosophy. Regents still consider these concepts as paramount in order to account for natural becoming. A possible reply to the criticism lies in the fact that moral and natural philosophy share some key notions because they share a common ground: that is, the structure of finite beings. Thus interpreted, moral and natural philosophy reflect the same nature under different aspects, in two distinct disciplines which are inevitably intertwined. It is true that Scholastic philosophy divided disciplines according to the method of enquiry proper to each; but also, the unity of a discipline is given by the unity of the subject. This shared terminology is not perceived as an illicit step, also because, I believe, regents had a strong awareness of the autonomy of natural philosophy.[18]

Stevenson implicitly holds that the appetite of prime matter is *appetitus perfectionis*, a formula that we also find in Reid 1610 and 1622. This result is obtained by proving that other sorts of qualification, such as *desiderij*, *complacentiae* or *privationis*, do not apply to the appetite of prime matter. 'Desire' is about an absent good, something missing and willed for insofar as missing: the fulfilment of the desire immediately removes the cause of desire, initiating a feeling of pleasure (*complacentia*) due to the enjoyment of the now present object of former desire. Privation on the contrary is the condition of absence of

[18] In part II I shall seek to offer more evidence for this claim by the examples of the careful ways in which regents treat the notions of final cause, of intelligence as the heavenly motor and of form as principles of falling bodies. I argue that regents can respond to the Moderns' criticism in a sound way.

good: in moral philosophy it is not a feeling internal to the moral agent, but indicates something missing. In natural philosophy, as we have seen, privation is the absence of a new form (and the presence of the actual form), and is posterior to the appetite. It is not absolute but relative absence: matter can never be without a form, so the absence is always relative to a form. None of these descriptions applies to appetite, which is the natural inclination of prime matter towards form.

Particularly interesting is the analysis of the differences between appetite and privation. Privation is the absence of the new form; matter is always in a state of privation, because it is essentially potency and no form can fully satisfy it. It is clear that appetite is different from privation, as a potency proper to matter from the state in which matter is. Appetite is prior according to reason because it is the potency of matter that permits us to talk of privation as a principle of becoming, not the opposite. Appetite is also prior according to reality, because appetite is more universal than privation, and what is more universal is always prior to what is less universal.

2. *Partim Eductiva*: prime matter and *eductio formae*

Eustachius identifies the second potency of prime matter in 'being eductive': *"ex materia possunt formae materiales virtute naturalium agentium educi"* (*SPhQ*, III, II, I, III). This concept introduces the aspect of prime matter regarded as the origin of forms, which integrates the notion of prime matter as receptive of forms. In fact, the two aspects are always conjoined, with one fundamental exception: the rational soul. In all cases but human beings, the matter of natural bodies is at the same time and in the same respect receptive and eductive, because it receives forms but also forms are coming out of, are taken out of matter (*e-ductae*). This is another case when Scholastic natural philosophy seems to rely on metaphors employed as technical terminology, as happens with 'appetite'.

Equally, the notion of 'eduction' does belong to natural philosophy, and it is the name of the process by which forms are immediately coming from matter and informing matter. The distinction between 'informing matter' and 'being educed from matter' is only logical: there is not a time when a form is first educed from matter and is then informing the same matter, or vice versa. The distinction is then one of reason, but it has a foundation in things, because the two terms actually refer to two distinct aspects of the same process. Rational souls are exceptions, as already noted. For a form to be educed from matter, as Eustachius observed, it is required that form is material (*materiata*), so endowed with the corruptibility proper of material things. Rational soul is thus not educed from matter, because this would be a direct argument for its mortality; the rational soul is created at the very moment of the information of the matter of the newly conceived man. This happens by direct intervention of God, which compensates for the inadequacy of the material world for originating a human being. This act of God is an act of his *potentia ordinata*, as it does not take place above nature *(supra naturam)* or against nature (*contra naturam*) by absolute powers, but within nature. Concurring in the generation of men is one of the ordinary means by which God continuously keeps the created world perfect.

The common formula we find in the theses has it that:

> licet forma educatur, non ex tamen materia gignitur. [Young 1613, TP 2.III][19]

There is then a difference between 'being educed from' matter and 'being born from', 'being brought forth from' matter (*gignitur*). In the second case in fact, being born entails a dependency of form on matter which is unacceptable, because it would call in question the theory of form and matter as 'principles' of the composite. A principle originating from another principle would not be a principle anymore, for it would depend on something else, as Aristotle explains in *Phys.* I, 6. Form and matter must be preserved in their opposition as contraries, neither depending on anything else. Scholasticism reinterpreted this theory as well in the light of the Christian faith, making form and matter still mutually independent as principles, but ultimately dependent on God as first principle insofar as they are created. On the physical level though, the Aristotelian theory remains unchanged.

We can distinguish in the theses two accounts of eduction: the first shared by the vast majority of regents, the second held by just one of them, J. Dalrymple, regent at Glasgow University, and author of the only set of graduation theses from Glasgow University in the first half of the seventeenth century. He is bringing forward a noteworthy theory of the direct intervention of God in the eduction of forms.

2.1 Traditional theory of *eductio* in the *Theses*

The most interesting passage is in Fairley 1619. The regent's conclusion is:

[19] On eduction see also, for example: Fairley 1619, TP II; Reid 1622, TP II; Fairley 1623, TP I, 3-4; and Martin 1618, TP XVII and King 1624, TP V, in particular in relation to the generation of a human being.

> ergo et formam educi e potentia materiae duo postulat: 1. Ut forma fiat in actu, cum prius solum esset in potentia: 2. Ut sine materiae adminiculo nec effici nec permanere possit. [TP II.4]

The connection between form and matter acquires its full evidence and depth in the process of eduction. Form and matter can initiate a compound which is one per se (including the rational soul and matter), which already indicates an essential unity. An even stronger claim is made when form is said to come from matter. In no way is form an accident of matter, or a mode of matter: regents hold onto the real distinction between form and matter as distinct principles. Eduction should not be misinterpreted as a derivation of form from matter, as regents warn with the words *"ex materia tamen non gignitur"*. According to Fairley, for a form to be educed from matter two things are required: 1) a passage from potency to act; 2) a dependence on matter limited to the mode of being, in virtue of which matter is prior to form. These two requirements explain eduction but also serve as a principle of distinction between *formae materiales* and *formae immateriales*: the second requirement does not apply to the rational soul, which can exist without its compound.

A material form is such because it is educed from matter. Matter here gets as close as possible in Scholastic natural philosophy to some sort of activity, which is never attributed to matter. Even the grammar of Fairley's sentences is revealing: the regent uses form as subject of a passive verbal form, and does not use matter as subject of an active one. Eduction is then something happening to form, not something caused by matter to form. Matter is a necessary component of the process (material cause), but is not the efficient cause of the existence of form. It is not a contradiction that form is materially caused by its contrary, matter: in fact, form is dependent on matter in exactly the aspect that is proper to matter, materiality, not in its own, formality.

Fairley's account includes the rejection of two objections raised against the theory of eduction: 1) that eduction implies that forms pre-exist in matter in order to be able to be educed from it; 2) that forms are created in matter.

> Si formae materiales nullo modo praecederent in materia, sed tantum in potentia activa agentis, crearentur. Hinc I. formae materiales praecedunt in potentia materiae. 2. Esse in potentia materiae est praecedere potentiam materiae a qua forma nata est dependere in Fieri et Esse. [TP II.1-3]

Forms do not pre-exist in the potency of matter because existence entails being actual from the side of forms: an actually existing form informs matter, and hence there is no room for eduction. The word chosen is 'to precede' (*praecedunt*): forms are in potency within matter, in the sense that matter is potentially informed by forms, not that forms are in potency to existence already within matter. This would lead to the coexistence of infinite potentially existing forms within matter, regarded as an absurd conclusion.

By this claim, the objection of the continuous creation of forms in matter is rejected: if forms were just dependent on the active virtue of the agent, then they would be created in matter: but Fairley holds that material forms do originate from matter. So, in order to bring it about that there are material forms, three elements are required: 1) a material cause, matter; 2) a formal cause (a material form preceding in the potency of matter); and finally 3) an agent, an efficient cause activating the preceding form and causing it to inform matter. This agent is identified in any other natural being acting on matter. It is the adequate physical cause for the eduction of form, because it alone is sufficient for form to be educed. It is not necessary for it to be the primary cause: in fact, an instrumental cause is enough. Instrumental causes are causes directly affecting something else, not by their own powers, but by the powers of the primary cause, by which they are used. An instrumental cause can thus be a real cause, even if it is not a primary cause. This is how Baron defines it:

> *Instrumentalis* vero, ut loquuntur Scholastici, est quae ab alio agente elevatur ad effectum producendum, quem non potest producere propria sua virtute. [*Metaphysica generalis*, sectio VIII]

The case of the rational soul is again illustrative: no natural body can be the primary cause for the birth of a man, because God's intervention is always required. In the natural course of events though, God requires an instrumental cause in place, not because his absolute power alone could not create a man, but because God's intervention is inserted in the natural process of procreation. It is a principle in Scholastic natural philosophy that finite beings are endowed with actual powers of their own, so that they can act as primary causes and exert a real efficient causality. It is precisely against this theory that Dalrymple formulates his objection and his alternative solution.

Before moving to Dalrymple, one last remark is important. Rankine 1627 introduces the notion of inherence in the analysis of eduction. Material forms are educed by an agent from matter, which is receptive of form and at the same time is acted upon by the agent. In order to account for the essential unity of the newly former compound, the question about the identity between eduction of form and inherence of form must be addressed:

> Forma materialis non habet propriam subsistentiam sed inhaeret materiae. Ergo per quam actionem educitur de potentia materiae, per eandem materiae inhaeret. Et cum sit eadem actio, sit inhaesio formae in materia et unitio ejusdem cum materia (non enim potest inhaerere materiae, nisi uniatur) per quam actionem educitur de potentia materiae, per eandem ei unietur. [TP IV]

Eduction and inherence are the same action, there is no real distinction. The three moments of eduction, inherence and information of matter are temporally one and are only logically distinct.

2.2 Dalrymple 1646: criticism of regents on *eductio*[20]

Dalrymple structures his criticism of the traditional position of regents on eduction on the basis of his low opinion of the potencies of matter. His set of theses is very interesting, first, as I mentioned, because it is the only existing set from Glasgow in the first half of the seventeenth century; secondly, because of the feeling of a breaking down of Scholastic philosophy that we get from his pages. It is possible that Dalrymple was more responsive than other regents to the challenges to Scholastic philosophy raised by the new philosophy. It is arguable that his set of theses represents an early Scottish attempt to incorporate themes of the 'new philosophy' within the body of the established Scholastic teaching in the universities. It is regrettable that no other sets of theses from the same period in Glasgow are available, for this limits our ability to judge the actual novelty of Dalrymple's philosophy.

His eclectism is well represented by his theory of eduction. He comes close to rejecting the whole notion as unintelligible:

> Originem et productionem formarum ascribere eductioni de potentia materiae, inextricabile latibulum, cum potentia materiae omnino inefficax, sit tantum passiva et receptiva, atque eductio saepe fiat per causam Instrumentalem, aut inferiorem effectu

[20] A translation of the *Theses physicae* of Dalrymple 1646 and biographical information are in the Appendix.

> producendo. Productionem formarum nos DEO ascribimus, propagationem vero ejusdem formae productae unioni. [TP X-XI]

I believe that Dalrymple brings about an interesting shift in the meaning of the concept of potency of prime matter, which makes the traditional reading of eduction unsustainable. He opens his passage by stating the difficulty of the subject (*inextricabile latibulum*), due to (*cum*) the ineffectiveness of the potency of prime matter to perform the eduction of forms. Prime matter's potency is only passive and receptive, and must always be supported by an instrumental cause. Prime matter is here understood as a physical cause lacking the sufficient power (*potentia*) to perform: it is exactly this sense of prime matter which is in contrast with the Scholastic notion. Prime matter is a material cause, which by definition is the material principle of the compound. It is not uncommon to read about the ineffectiveness of prime matter, but its ineffectiveness is always related to prime matter's metaphysical act being insufficient to grant independent existence. Dalrymple transfers this ineffectiveness to the sphere of natural causality, shifting from the metaphysical to the physical level. Furthermore, prime matter and its potency are treated in general as 'causes', without the due qualification of 'material'.

In a standard Scholastic doctrine, the fact that potency is *"tantum passiva et receptiva"* is never seen as a limitation of matter's role in the origin of the compound, but it is precisely the role of prime matter as physical principle. Dalrymple seems to take 'tantum' in the sense of 'just', 'merely', thus implying a weak causality unable to cause on its own. I argue that this specific theory implies some sort of rejection of the idea that finite beings are true causal agents.

The unsatisfactory potency of matter is thus compensated by a direct act of God: *"productionem formarum nos DEO ascribimus"* (*ibidem*). This is not a Scholastic doctrine stricto sensu, even if it is still formulated in Scholastic terminology. The doctrine of the autonomy of created beings in the natural world is always maintained by the regents, who are keen not to postulate God's intervention. Ultimately, all of reality is dependent on God, who is the first or primary cause: it is possible to say, when holding the theological doctrine of analogical predication, that God alone is a true cause and consequently is the only cause. Yet, in Scholasticism this discourse never led philosophers to deny that *within* a context of natural philosophy it is correct to ascribe real causality to creatures. This is what Dalrymple seems to claim: if natural substances are not the primary causes of the production of material substances (because prime matter's potency fails to educe material forms), then natural substances are deprived of physical causality, and are just instrumental causes.

I believe that a comparison with a standard account of the activity of secondary causes in the theses can shed light on Dalrymple's own position, and in general on the afore-mentioned 'autonomy' of natural philosophy as a discipline. Forbes 1624[21] deals with this question in the metaphysical section of his *Theses philosophicae*, III-V. The regent expounds two opposite views: 1) that causal efficiency does not belong to secondary causes, a theory ascribed to the 'Arabs'; and 2) that the created substances are alone enough to bring about their effects, a theory ascribed to Durandus and his followers. [TM III]. Both theories are regarded as absurd and dangerous for philosophy and faith. In the former case, Forbes believes that the contingency of things and the freedom of our will would be annihilated, because God would be the only true cause of natural monsters and of our sinful behaviour. This is not all: these consequences are not less important than the fact that this theory:

> scientias destruat, rerum quidditates et facultates, in occulto naturae recessu abscondens, et communissima evertat axiomata, qualia: Sol illuminat: Ignis calefaciat. [TM IIII]

If there is no real secondary efficient causation, natural philosophy as a science is in danger. The second theory is no less false, since it overturns the natural order of beings, and the nature of created substances, which always needs the concourse of the first cause. Forbes's answer to the dilemma seeks to include dependence on the first cause and true efficient causality in the nature of created substances:

> Ita quicquid entitatis in operationibus est, id essentialiter a DEO pendet, et a summo Ente [...] dirigi, et in finem ordinari necesse. Potest quidem causa secunda, exclusis aliis ejusdem generis, simile sibi producere. [TM V]

The power and presence of God is the same in respect of the action of the creature: *"ut virtute et praesentia eadem etiam qua creatura actione"* (*ibidem*).

---

[21] John Forbes of Corse (1593-1648), minister of the Church of Scotland, theologian, regent and member of the Aberdeen Doctors. Entered King's College, Aberdeen, in 1607 and probably graduated in 1611. We have no graduation theses for that year. Forbes started a tour of European universities in 1612, which brought him first to Heidelberg, then to Sedan (1615), where he studied with Andrew Melville. Ordained in Aberdeen in 1620. He wrote the graduation theses for 1624 at King's College. He refused to sign the National Covenant, and continued to act in support of episcopacy and of his own religious convictions. He was eventually forced to leave first his academic position in 1641, then Scotland in 1643. Died in 1648 after returning to Aberdeen. One of the main figures among the Doctors, Forbes represents well the independent spirit of Aberdeen in matters of religion and ecclesiastical organisation. *DNB*.

Sibbald 1625 agrees with Forbes: he points out the contradiction between the freedom of our will and predermination, if all causality is from God. Therefore:

> Concursus DEI et actio causae secundae sunt una eademque numero actio. [...] effectum creaturae dici a concursu DEI pendere, actionem vero non item, cum actio et concursu DEI sunt idem, idem autem non pendet a seipso. [TM VII-VIII]

It seems that Dalrymple disagrees with the other regents on eduction and causality of secondary causes. His theories then prompt a question about his sources: by 1646 it is likely that, as an educated member of a distinguished family in Glasgow, Dalrymple had become acquainted with the most recent novelties in philosophy, either by travelling or by having access to books locally, sometimes even before the university had bought them. As likely as this sounds, I am reluctant on the basis of the historical evidence at our disposal to support this claim. There is another passage by Dalrymple which again seems to break away from the Scholastic tradition:

> Toti materiae massae unam et intimam formam corporis DEUS in principio impressit, unde constituatur in ratione corporis, quaeque jam in omnibus manet eadem, nec contrariam habet unde expellatur, sed materiae coaeva est, et coaetanea. [TP XII]

Dalrymple is very clear: God impressed an intimate, coeval, inseparable and unique form upon the whole of matter, by which it is constituted as body (*in ratione corporis*). The regent chooses to transform the traditional notion of prime matter into the notion of a body, essentially informed by direct act of God; thus Dalrymple is in opposition to the other regents. In this theory there are elements which resemble Descartes' notion of matter as res extensa; or, alternatively, Zabarella's of matter as body. Archival evidence shows that Zabarella's works were held by Scottish universities, and his name is often mentioned in many theses; yet, Descartes seems a likelier source. To support my view, I wish to mention the opinion on Dalrymple by Skene, regent in Aberdeen in the 1680s, in his *Positiones aliquot philosophicae*, Aberdeen 1688:

> Sola cogitatio menti tribuenda est, ut extensum ad corpus, ita est et cogitans ad mentem. Substantia est immortalis, et immaterialis, cui repugnat existentia in loco. Rationem spiritus formalem posuit D. *De Stair* in perceptione. [VI.15]

Skene's set of graduation theses expounds the most important philosophical schools: the longest sections are on *philosophia peripatetica* and *Cartesii philosophia*. The regent offers a historical analysis, to my knowledge unique in the Scottish universities, which might perhaps be regarded as an early work in history of philosophy. I shall return to it in the Conclusions, section 2.2. Dalrymple, later Viscount Stair, is the second most quoted authority after Descartes in the section on Cartesian philosophy. There is evidence that Skene regarded Dalrymple (probably basing himself on his *Physiologia nova experimentalis*, Leiden 1686), as if not a Cartesian, at least as a 'new philosopher'. On the evidence of Dalrymple's graduation theses and Skene's interpretation, it is then arguable that Dalrymple had been investigating modern philosophy ever since his regenting years in Glasgow. This would explain why he is the most critical regent of Scholastic natural philosophy in the first half of the seventeenth century.

3. Conclusion

Following Eustachius's analysis of the potencies of prime matter, I have structured this chapter in two parts: the first one on prime matter as receptive principle, the second one on prime matter as eductive principle.

Receptiveness and eductiveness are the two potencies of prime matter. Qua potencies, they flow from the essence of prime matter (investigated in chapter 1) even if they are not included in the definition of the essence of prime matter. The analysis of such potencies is thus the first step into the analysis of prime matter as principle of the compound, and not simply as a metaphysical principle.

The potencies of prime matter imply the relationship with form: all forms are either received by prime matter or educed from it. The first aspect of this relationship is that prime matter is receptive of forms: prime matter has an appetite towards form, which is the 'good' and the end of prime matter. I have investigated the case of the rational soul as the example of a form which is independent from matter: the rational soul is received by prime matter, and not educed from it. This debate will be completed in the next chapter with the analysis of the bodily form.

The second aspect is that prime matter is also eductive with respect to form; that means that material forms are educed ('taken out of') matter in virtue of a number of causes:

matter as the material form, the new form as the formal cause and an external agent acting on the material cause as the efficient cause.

I have then investigated the set of theses by Dalrymple 1646. The regent puts forward an interpretation of the potency of matter and of the causality of secondary causes which seems to break with the Scholastic tradition in natural philosophy.

De proprietatibus materiae primae

Eustachius's words can serve us well when introducing the analysis of the properties (*proprietates*) of prime matter as well. In pars III, tractatus I, disputatio I, quaestio IV of his *Summa philosophiae quadripartita* we are told about four properties:

> prima est, Quod sit quanta. Adeo enim materiae propria est quantitas, ut ipsi primo et per se competat; deinde per ipsam toti composito naturali. Adde etiam, formam, sive substantialem sive accidentalem, non nisi mediante quantitate in materiam recipi. [...]
>
> Secunda est, Quod sit ingenerabilis et incorruptibilis; licet *mutabilis* dici possit quatenus mutationum vicissitudines experitur, dum succedentes sibi invicem formas suo sinu excipit. [...]
>
> Tertia est, Quod materia nunquam possit esse nuda. [...]
>
> Quarta proprietas est, Quod materia sit omnino passiva, i. e. nullam habeat potentiam activam sed tantum passiva.

Prime matter is thus endowed with four properties, in virtue of its essence: 1) being quantified; 2) being ungenerable and incorruptible; 3) being always informed; 4) being passive potency. The *Theses philosophicae* agree with Eustachius, whom I take here to be representative of the wide family of Scholastics, on this general account of prime matter's properties. This agreement though does not mean that Eustachius's explanation of these properties is the same as the regents'. Among the four types of properties, the one that is debated most is quantity, and it is from here that the regents part company with contemporary Scholasticism to build a theory compatible with their doctrine of the Eucharist. That doctrine is the topic of chapter 4, and part of the role of chapter 3 is to expound the crucial point that regents intended prime matter as essentially quantified; which point is the philosophical ground of the rejection of the Catholic dogma of Transubstantiation. The philosophical explanation of the dogma of Transubstantiation rests on the theory of the relation between substance and accident, of which the relation between prime matter and quantity is a case. The connection between the dogma and this philosophical theory is so strong that both Eustachius and Suárez feel compelled to mention the Eucharist when dealing with the properties of prime matter:

> una eademque materia variis sibi invicem subinde succedentibus formis subest; ita una eademque quantitas in illis perseveret; imo nonnunquam ipsius materiae vices gerat: ut contingit in augustissimo Eucharistiae sacramento. [Eustachius, *ibidem*]
>
> *Approbatur sententia reipsa distinguens quantitatem a substantia.* Atque haec sententia est omnino tenenda; quamquam enim non possit ratione naturali sufficienter demonstrari, tamen ex principiis theologiae convincitur esse vera, maxime propter mysterium Eucharistiae. [*DM*, 40, II]

In my exposition, I shall focus mainly on quantity as the key property of prime matter, working as the copula between the philosophical analysis of prime matter and the philosophical rejection of a theological dogma; or, in other words, between philosophy somehow restricted to the sphere of a purely intellectual enterprise and philosophy engaged with one of the main features of the epoch-making event of the Reformation. I argue that all philosophical doctrines held by regents regarding prime matter must be seen in the light of the broader context of the clash between different confessions of faith.

In this chapter, I intend to concentrate on prime matter still abstracting from the role that prime matter plays in the philosophical reading of this theological dogma. Also, I postpone the question of the priority of philosophy or theology in shaping the debate in the *Theses philosophicae*. This is an appropriate ordering because prime matter is first and above all a philosophical concept dealt with in a philosophical context: it is thus subject to analysis independent of any other discipline. Furthermore, the role of prime matter in the debate on Transubstantiation is relevant in proportion to its philosophical coherence and richness: in this sense, philosophy must be truly preparatory to theology.

Quantity, though having primacy, does not overshadow the remaining three properties. In the previous chapter prime matter has already been analysed with respect to being 'always informed' and 'passive potency': in this chapter, these two notions are going to be integrated into a more complete account of the role of matter in the compound. In fact, contrary to potencies of prime matter, properties are fully intelligible only when analysed in relation to form. The order of exposition follows the structure of prime matter, and this chapter is about the most specific of prime matter:[1] its relation to form and its role in the compound. In the previous two chapters, the analysis was still on the general level of the essence of prime matter and on prime matter as receptive and eductive potency.

On the basis of the analysis of the properties of prime matter, it is also possible to begin to form an account of the theory of substance, which helps to answer the question of what kind of Aristotelians the regents were. I shall focus on the reception of Aristotle in Conclusions, section 2. Given the importance of the notion of substance in any Scholastic philosophy, the account will have to be augmented by the analysis of movement in part II.

This chapter is divided into two sections: the first one is about the properties of ungenerability and incorruptibility of prime matter. An interesting theory is that of the resolution into prime matter:[2] when a physical compound becomes corrupted, resolution occurs if the remaining accidents inhere in prime matter, immediately without a form. In this theory, prime matter is a substratum of accidents, and its property of being the root of physical becoming is best explained. The second part deals with the relation between prime matter and quantity. Scholastics held that matter is quantified, in the general sense that a form obtains extension in space in virtue of its union with matter. This general theory does not suffice: it is important to investigate what sort of relation is established between matter and quantity, for example, addressing questions such as whether quantity is essentially extension in place, or merely extension of parts beyond parts;[3] or whether matter is really distinct from quantity. As I mentioned before, this account of quantity and matter cannot be fully understood without the reference to Transubstantiation.

---

[1] 'Most specific' is to be understood in the genus-species context: the analysis of the essence of prime matter is what is most general (like the genus); the analysis of the relation of prime matter with form is what is most particular (like the species).

[2] I have decided to translate the Latin formula *resolutio in materiam primam* with the English formula 'resolution into prime matter'. *Resolutio* is a technical Scholastic term: *"Resolutio est cuiusque rei ad sua principia, unde componitur, revocation: seu, est operis facti reductio ad principia, id est, ea, e quibus compositum est"*, (R. Goclenius, *Lexicon philosophicum*, Frankfurt 1613, art. Resolutio).

[3] Extension of 'parts beyond parts' (*partes extra partes*) means that an extended body has parts which are distinct among themselves by dimensions and mass. It is an extension *in ordine ad se*, which is a mode of a substance, not a mode of a quantity: therefore, it does not imply extension in place (space). Having parts beyond parts is a prerequisite to be extended in place. See, Ruvius, *In universam Aristotelis Dialecticam*, 1603, cap. 6, q. I and R. Goclenius, *Lexicon philosophicum*, art. Extensio.

1. Prime matter as incorruptible and ungenerated

Prime matter is a principle of compounds: as a principle, prime matter cannot depend on another principle, because principles are, by definition, the first components and explanations of something. Were prime matter explained by introducing another principle, it simply would not be a 'principle' any more. Likewise regarding form: form and prime matter are functional to one another; they are essentially open to one another, and thus depend on one another. It is not contradictory that two principles are mutually dependent: it is contradictory that one principle is explained by another principle. The analysis of the properties of prime matter is the analysis of prime matter as a principle in mutual dependence on form.

It must be pointed out that in a Christian metaphysics prime matter and form are principles per se of compounds only *secundum quid*, namely within the sphere of natural philosophy. In fact, they are principles ultimately depending on God, who alone is a principle per se absolutely speaking of any reality. This is a fundamental revision of the Aristotelian theory of substance, which allows for the acceptance of prime matter and form as principles per se, and does not admit a higher level of dependence. In fact, in Aristotle's philosophy there is no absolute efficient causality, and God (the prime motor) is the final cause of the universe. Scholastics differ from Aristotle not on the basis of a different definition of principle, but simply on a different application of the definition.

So, the two properties of 'incorruptibility' and 'ungenerability' follow from the definition of prime matter as principle, and have been introduced already. Prime matter is then an incorruptible and ungenerated principle of compounds: these properties are not included in the definition of prime matter as 'entitative act whose essence is being receptive pure potency', but they nonetheless flow from the essence. They belong to the definition of prime matter as a principle of the physical compound, not as a metaphysical act. When further analysed as principle of a compound, prime matter is essentially incorruptible and ungenerated.

Why is that so? In sum, regents explain this point implicitly during the quod sit analysis of prime matter. A demonstration of the existence of prime matter is obtained by means of the principle that nothing can come out of nothing combined with the rejection of the *regressus ad infinitum*. Things change, come-to-be and cease-to-be (*fieri* and *desinere*): in order to avoid an infinite regressus and a continuous creation of things from nothing, according to the Scholastics we are compelled to admit a first principle which underlies all these changes and makes them intelligible.[4] Aristotle claimed that the world and natural change are eternal, while Scholastics held that the world is created (that is, has a beginning in time), but they all argued on the basis of the rejection of the regressus and the acceptance of the principle that 'nothing' is not a principle. The properties of being 'incorruptible' and 'ungenerated' (regents do not offer any analysis of the primacy of one property over another) are thus essential properties (yet not part of the definition), because they are not demonstrated in the course of answering the 'utrum sit' or in the 'quid sit' questions, but rather they are presupposed by them. In Aristotelian fashion, a science does not yield the definition or pre-comprehension of its object, but enquires into an already 'given' object.[5]

When it comes to compounds, prime matter cannot be deprived of any of its essential properties; what forms do is to make prime matter formally actual and make it the matter of such and such a compound; they do not change the essence of prime matter in any way. Again, a principle does not change the opposite principle, it simply unites with it. Regents hold that prime matter is an incorruptible and ungenerated *component* of compounds, and in respect of the theory of natural substances this qualification carries weight in our understanding of its relation with form.

1.1 *Resolutio in materiam primam* and *forma mistionis*

All regents agree on the idea that prime matter is an entitative act, whose essence is being a receptive pure potency.[6] In other words, it is the purely receptive component of compounds. It is also the incorruptible and ungenerable purely receptive component of compounds: prime matter is a 'something' cooperating in the compounds by being receptive, incorruptible and ungenerable/ungenerated.[7] These properties belong not only to prime matter considered as a principle, but also to prime matter considered as a component. That is, the actual, individualised matter of any compound is incorruptible and ungenerable, not just prime matter intended as a metaphysical principle. The theory of prime matter as entitative act within the framework of a metaphysics of essence is the basis

[4] This is the backbone of the proof from natural philosophy as we have seen it in St Andrews 1629, part I, chapter 1, section 2.1.

[5] J.-F. Courtine, *Suarez et le système de la métaphysique*, Paris, PUF, 1990, p. 19.

[6] As shown in chapter 1. See also: Fairley 1615, TP VI.4-6; Forbes 1623, TP I; Stevenson 1629, TP VIII.3; Barclay 1631, TP I.3 Wemys 1631, TP I.

[7] If something is not in potency towards being generable, it follows that it is ungenerated.

for the regents' analysis of matter as component and the essence of prime matter does not change whether we consider it abstracting from its union with form or not.

The theory of resolution into prime matter (*resolutio in materiam primam*) that we find in the theses is closely related to the properties of incorruptibility and ungenerability. In Scholasticism, the debate concerns whether forms inhere in matter immediately or mediately: whether a non-substantial form needs a substantial form in which it inheres immediately and in virtue of which it inheres in matter mediately. In other words, whether prime matter can be the subject of non-substantial (accidental) forms, or not. Many proponents of the doctrine of a mediated inherence of accidental forms in matter through a substantial form reject the doctrine of resolution into prime matter. Thomas Aquinas is one: according to Thomas, all accidents inhere in a substantial form immediately, and in prime matter mediately. A corollary of this theory is that there is only one substantial form for each compound. Regents on the contrary take the side of Scotus, who holds that there is a plurality of forms in a compound.[8]

We shall see that some regents, while accepting the Scotistic framework and the concept of bodily form, do hold that even accidental forms can inhere immediately in prime matter. The qualification of 'accidental' is important: there is no doubt regarding the immediate inherence of substantial forms. Substantial forms are the forms which alone originate a compound (like the rational soul in the case of men), while accidental forms are the forms of the accidents which qualify a compound (like the colour of the hair of a man). Substantial forms originate a substance (category 1), accidental forms originate accidents, the categories of quantity and quality, regarded as the two categories on which all the remaining seven categories depend. The question is thus whether accidental forms can qualify a compound which is not already qualified by a substantial form.

Following Scotus, the regents distinguish between animate and inanimate beings. Reid 1614 holds that:

> Viventia non resolvuntur in materiam primam; at non viventia resolvuntur omnia. [TP 24.2]

and concludes that:

> non in omni corruptione resolutio fit in Materiam primam immediate. [TP 24.3]

[8] J. Duns Scotus, *Ordinatio*, IV, d. 11, q. 3, n. 45.

In the first passage Reid holds that animate beings do not resolve into prime matter; while inanimate beings do resolve into prime matter. In the second one it is claimed that such resolution does not occur immediately in all corruptions. Both quotes are the conclusions of longer passages.

1.1.1 *Resolutio* and animate beings

It seems that the difference lies in what sort of compound Reid is talking about. In the corruption of animate beings, no resolution takes place because, as we read in the majority of regents, including Reid, there is something added to the substantial form-matter relation, some sort of medium, which is missing from inanimate beings. The most apparent difference is that animate beings, by definition, have a soul (vegetative, animal or rational). Yet, Reid does not have this in mind when rejecting the doctrine of the resolution into prime matter: his reference is to the form of mixture (*forma mistionis*)[9], which is defined by Baron 1627 as follows:

> Forma mistionis non est viventium forma generica nec ullum ijs essentiae gradum tribuit, sed constituit mistum illud incompletum quod est altera essentiae pars physica, et corpus viventis appellatur. [TP VIII.4]

A definition which can now be coupled with the longer passage in Reid:

> Forma mistionis non est superaddita formis elementorum; sed Anima formae mistionis vere superadditur. Sublata Anima potest remanere mistum, at sublata forma mistionis, praeter Materiam primam nihil supponitur. Viventia non resolvuntur in materiam primam; at non viventia resolvuntur omnia. [TP 24]

Baron and Reid agree on the notion of a form of mixture. Baron points out that: 1) this form is not the generic form of animate beings (generic form 'man' when talking about a single man) because regents hold that no generic being can exist, but only individuals; 2) it does not confer any degree (*essentiae gradum*) to the essence of animate beings, so it does

[9] I shall translate *forma mistionis* with 'form of mixture', in the sense of a form 'based on mixture', even if it might be open to misinterpretation. A possible alternative translation is 'form of compound', which I already use to translate *forma compositi*.

not follow from their form; finally that 3) it is 'the other physical part of the essence' (*altera essentiae pars physica*), which we can call 'body'.

Reid's account sheds some light on the relation between form of mixture and corruption. Soul in general (and therefore including all animate beings, not humans only) is said to be added (*superadditur*) to the form of mixture, in such a relation that: 1) the soul can be in a compound only posterior to the presence of the form of mixture; and that 2) the corruption of the soul does not entail the corruption of this form. The relation is clearly not that of identity, because the latter can be without the former, yet not conversely. And as a final remark, only once the form of mixture is corrupted can there be a resolution into prime matter, because there is no medium between this latter form and prime matter. What immediately inheres in prime matter is thus the form of mixture, not the soul. Therefore animate beings do not resolve into prime matter immediately when they corrupt (when the unity between the soul and the body corrupts), because the form of mixture remains. This is not the case of inanimate beings, which have no form beyond the form of mixture.

Putting the two passages together, this is the general account of compounds that the two regents hold: 1) souls need the form of mixture in order to inhere in or inform matter; 2) what they need is matter already constituted as a body, in virtue of the form of mixture; 3) this form is thus present in any physical compound, and immediately inheres in prime matter.[10]

Two words in these accounts should not pass unnoticed. First, the reference to 'body' in Baron 1627: it is not a novelty in Aristotelian philosophy, as Zabarella had previously held that matter constitutes itself immediately as body; yet, it is not a commonly accepted Scholastic doctrine. Regents claim something different though: it is not matter alone which can be called 'body', but matter when informed by the form of mixture. In order to obtain a body some form (some ordering of the underlying matter) must be provided. This ordering is not posterior to the soul and caused by it, but prior to it and necessary in order to have a soul informing a compound. Is this the form educed from matter? Baron and Reid do not make such a connection for us, but it is arguable that, with the exception of the rational soul, all forms, both souls and forms of mixture, are educed from matter. The form of mixture is a material form, therefore it is educed from matter.

Form of mixture seems to be an unnecessary third element added to the structure of compounds, which could be intelligible with only two elements in play, matter and form for inanimate beings and matter and soul (= substantial form) for animate beings; indeed, this is Thomas's theory. Following Scotus, the regents introduce this third element in order

---

[10] See also, Sibbald 1625, TP I.

to account for the empirical evidence of the preservation of the body of animate beings through the process of corruption, or better, to preserve the numerical identity between the body and the corpse of animate beings.[11] It seems evident that we can identify the corpse of Socrates by its identity in appearance with the former living body of Socrates: the traditional example is the numerical identity of a scar on the corpse and on the body. We can say, regents argue, that the scar is the same; we can even say on this basis that the corpse is the corpse of Socrates by means of the physical identity with Socrates before death. There must then be something more solid than just resemblance if we are to formulate a judgment of identity. The preservation of the ordered bulk of matter that we call the corpse of Socrates is thus due to the preservation of the form of mixture of the body of Socrates, a form which is not corrupted in the very moment of Socrates' death.

The second remarkable element is the terminological shift from 'substantial form' to 'soul': substantial form is virtually missing from these accounts, perhaps because it is too general a concept, and does not provide any explanation for the problem of resolution. The distinction between substantial form and accidental form is not in question; what regents do is to go beyond the identity between substantial form and soul when it comes to animate bodies. In fact, it is arguable that the form of mixture in the corpse of Socrates is the substantial form of the corpse. In principle, the objections that 1) the corpse does not act as a single unified body (= it is not alive); or that 2) it is not a stable compound, because it quickly corrupts, do not prove the theory false, because point (1) is applicable to any inanimate body, and (2) is proper to both animate and inanimate bodies. Regents intentionally speak of 'soul' to clearly mark the difference between what makes a body alive and what makes a body such.

1.1.2 *Resolutio* and corruption in general

Reid 1614's second quote is:

> non in omni corruptione resolutio fit in Materiam primam immediate. [TP 24.3]

which is the conclusion of the following passage:

---

[11] Duns Scotus, *Ordinatio*, IV, d. 11, q. 3, n. 45.

> Materia prima non est corpus sensile. An non ergo aliquid et per se quantum, insensile tamen erit. Non omnis quantitas est per se sensilis, nisi terminos habuerit. *Arist.* igitur corpus sensile tantum dicetur, quod actuatum est, et forma aliqua praeditum. Ideo materia prima sola, proprium est generationis subjectum idem sub utroque termino. Respectu subjecti unius et ejusdem sub utroque termino, non in omni Corruptione resolutio fit in Materiam primam immediate. [TP 23-24]

This passage is quite complex. It touches on a few fundamental theories, and the conclusion rests on the not immediately clear qualification *"respectu subjecti unius et ejusdem sub utroque termino"*. The qualification has to be explained in order to understand the conclusion. Reid accepts the Aristotelian doctrine that prime matter is not a sensible body, because only a defined quantity (that is, with termini) can be called 'sensible'. Thus, prime matter is sensible only when its quantity is given certain boundaries by form, and this only happens in a compound. The notion of sensible body falls under this description. The second part is more interesting: the regent introduces it by *'ideo'* (therefore), but the *sequitur* is not too clear. Reid appears to be saying that prime matter is sensible only when informed; and therefore only prime matter can be the proper subject of generation identical *sub utroque termino*, with the termini of generation being the initial moment (*terminus a quo*) and the final moment (*terminus ad quem*). Prime matter can be such a subject because in itself it has no termini; it can receive them only from form. Thus, with respect to the same individual subject (the subject undergoing change), and with respect to both termini (*a quo* and *ad quem*), it appears that resolution into prime matter does not occur in all corruptions (where 'corruption' here is taken to mean 'loss of all boundaries' and 'acquisition of new boundaries'). It is then explained again why in the case of corruption of animate bodies resolution into prime matter does not occur: there is no such a thing as 'loss of all boundaries' since the form of mixture remains.

1.2 Rejection of form of mixture: different theory of *resolutio*

The passages quoted above are representative of a tendency among the regents, who usually accept the following central points: 1) there is such a thing as a form of mixture; 2) matter informed by it is constituted as body; 3) resolution into prime matter occurs when the totality of a compound is corrupted: in the case of animate bodies, the corruption of the union between soul and body leads only mediately to resolution into prime matter, after the logically and metaphysically posterior corruption of the union between the form of mixture and the body; in the case of inanimate bodies, resolution into prime matter occurs immediately when the union between the only substantial form of the compound (form of mixture) and the body is corrupted; 4) there are thus two substantial forms within each animate compound, a soul and a form of mixture, and only one in each inanimate compound.[12] We can also argue that regents include the form of mixture in the number of material forms educed from matter.

There are nevertheless some regents who hold a different view on this subject, and contrary to the case of Dalrymple on secondary causality and the potency of matter, these alternative opinions do inscribe themselves within a more established Scholastic tradition. One case is Rankine 1627, who explains his view in a thesis under the heading: *'Materia prius respicit formas substantiales, postea accidentales, 7.* Metaph. text. *8.'* The passage is quite informative on some regents' rejection of the idea that the scar in a corpse is the same as the one in the formerly living body. His conclusion runs as follows:

> Non igitur manet eadem numero cicatrix in cadavere, quae prius fuit in vivente, licet sensus ita manet, cum sensus circa obiectum commune (cuiusmodi est unitas aut diversitas numerica) etiam debite approximatum errare possit. [TP XIV.7]

This passage is not the explanation why the scar is numerically different; Rankine is simply starting from the theory that senses can be wrong when apprehending a common object. In other words, senses are wrong when providing our intellect with the evidence of the resemblance between these two scars, which is then interpreted as the sign of the numerical identity of the scar. In Rankine, as much as in the other regents, the question is about the 'numerical' identity of the scar because the scar of the dead body does look like the scar of the living body. What differs is the type of identity. Rankine's explanation is to be found in the previous lines. He agrees with the idea that accidental forms are in matter only in virtue of the substantial form they inhere in: substantial forms are not required by matter in order to be a material cause (matter is receptive by essence), yet they are required for matter to be receptive as a material cause of accidents: matter is receptive towards substantial forms, which enables matter to also receive accidental forms:

[12] In the passages analysed, regents favour the expression *forma mistionis*: I believe that a perhaps more common expression for the same concept is *forma corporis*. Regarding the four points listed here, besides the texts already quoted, see also: points 1-2: Wemys 1612, TP 4; Baron 1617, TM II-III; Baron 1627, TP

> Omnia igitur accidentia quae in materia generantur, praesupponunt in materia formas substantiales, per quas materia redditur ens actu, atque ita idoneum subiectum accidentium.
> Forma substantialis licet ad materiam non requiratur tanquam concausa receptionis passivae in eodemmet genere causae, necessario tamen requiritur tanquam causa formalis, per quam habilis redditur ad sustentanda accidentia quae in eo generantur. [TP XIV.1-2]

The conclusion follows:

> Unio igitur substantialis, causa est unionis accidentalis. Ea igitur dissoluta, et altera dissolvetur necessario. [TP XIV.3-4]

This is why the scar cannot be numerically one: the numerical identity of the compound is dissolved the very moment the compound corrupts. The accidental form of the scar inheres in matter only in virtue of the substantial form: when Socrates dies, his compound dissolves (his substantial form parts from his matter), so the remaining scar cannot be the same scar, contrary to empirical evidence, as other regents would say. Rankine does not go further in his analysis. Rankine seems to reject the account of the form of mixture, with all its consequences. In particular, he seems to intend 'substantial form' as the unique form of a compound (with the exception of accidental forms). It is then hard to say whether Rankine can be counted as belonging to a Scotistic approach regarding this subject. One solution might be that Rankine includes the form of mixture in the general expression of 'substantial form': in that case, his theory would agree with that of Reid and Baron. Unfortunately Rankine does not clarify this point, so what his solution was is left open.

On the more general level of the definition and analysis of the essence of prime matter, regents show a vast agreement; in the more particular account of powers and properties, however, some differences among them become apparent. This is hardly surprising: within the same metaphysics of prime matter as entitative act several theories of the structure of compounds are equally available and coherent. This is the case of the form of mixture: we cannot say that this theory represents the totality of the theses, because an equally valid tendency is to account for the corruption of a compound with the presupposition of the unicity of substantial form. Unanimity is reached again with respect to the rejection of

VIII.4; Murray 1628, TP XI; points 3-4: Craig 1599, TP 10; Baron 1621, Disputatio physica, I; Sibbald 1625, De pluralitate formarum in eodem composito, TP I-V; Leech 1633, TP IX.

Transubstantiation, since it is not only a question of philosophical debate but primarily of confession of faith. As it appears, regents were then given autonomy in matters of philosophy; there was significant disagreement among them, and in the records of universities no mention of philosophical impositions can be found. The term 'Scotistic Eclectism' appears to describe the overall character of the *Theses philosophicae* quite accurately; but we are confronted with quite a number of regents in six different colleges across all Scotland, and perfect agreement among them is in any case unlikely. Their substantial acceptance, in general, of Scotism in natural philosophy[13] explains the remarkable fact that it is possible to treat the theses as a unified corpus of philosophical teaching, and not just as a corpus of philosophy that is, in some sense, "Scholastic".

2. Prime matter and quantity

Quantity is a fundamental property of prime matter. Chapter 4 will deal with the debate over the relationship between quantity, accidents and place. In this section the focus is on quantity as a property, and especially on the relation between quantity and prime matter with respect to the compound.

As a property, quantity is not part of the definition of prime matter: rather, qua *proprium*, it is an attribute possessed in virtue of the essence of prime matter. In the *Isagoges*, Porphyry defines 'proprium' in four different ways. The last one applies to quantity in relation to prime matter:

> 'fourthly, what belongs to the totality of a species always and exclusively, like, for example, the ability to laugh belongs to a man.' [12, 17-18][14]

A standard reception of this theory is found in Baron 1627, who writes that it is not more possible to separate the ability to laugh from the human nature than quantity from matter [TP III]. This passage will be relevant in the next chapter as a counterargument against the

[13] We have seen so far that the regents accept, in general, these central doctrines of Scotism: 1) metaphysics of essence, which includes 2) prime matter understood as, in some sense, actual; 3) the form of a body, which informs prime matter for the reception of the rational soul, which implies 4) the plurality of forms in a human compound. We will see in part II how Scotism also shapes the theory of movement of the regents, even if it carries less weight than in metaphysics.

[14] '[T]έταρπον δέ, ἐφ'οὗ συνδεδράμηκεν τὸ μόνω καὶ παντὶ καὶ ἀεί, ὡς τῷ ἀνθρώπῳ τὸ γελαστικόν.'

theory of the separability of quantity from matter that Catholic Scholastics bring forward when justifying the miracle of Transubstantiation. When dealing with Transubstantiation, regents focus on the analysis of the relation between matter and quantity, deploying a precise criticism of the view held by Catholic Scholastics. Apart from this context, the account of matter and quantity is usually centred upon what sort of contribution to the compound is proper to matter in virtue of its quantified nature. The focus on the compound as a substance is more evident here.

As early as Stevenson 1596 (the first set of theses available),[15] regents have it that form receives quantity from matter: this means that form, which is per se immaterial and indivisible, in virtue of the union with matter is made material and divisible. This is still a general statement, but sufficient to establish a logical and metaphysical tie between form and quality on the one side and matter and quantity on the other. Some regents claim that form and matter are two incomplete substances (category 1) from which respectively quality and quantity (categories 2 and 3) follow, somehow putting form and matter on the same level as subjects of accidents.

Three points seem to be involved here: 1) quantity as primarily related to prime matter rather than to form; 2) matter as a subject of accidents; and 3) the question whether this relation of quantity to matter weakens the substantial unity of the compound. This latter point finally introduces the debate on the kind of unity that is proper to physical substances. We have seen that the majority of the regents accepts the notion of the 'form of mixture', drawn from the Scotistic tradition. Medieval Scholastics divided themselves most famously between Thomists and Augustinians on this topic: according to the Thomists the substantial form is unique to a compound and the plurality of substantial forms endangers the essential unity of the compound because a unity per se cannot be the result of the union of two acts, namely the soul and the form of mixture. It appears that this question mainly concerns the account of the unity of the human substance, and the related status of the body. We will see how the regents are not unanimous in their theory of the union of the compound, even if they seek to establish a unity per se. This debate will be central in modern philosophy as well, and will originate from the Cartesian account of matter as res extensa.

[15] A translation of the natural philosophy section of Robertson 1596 is in the Appendix.

2.1 Relation between prime matter and quantity

Claiming that quantity is primarily related to prime matter does not tell us anything about the specific nature of this relation: within the theory of quantity as a *proprium* of matter, more than one direction is coherent with the premise. The standard Scholastic solution is that quantity, as an accident of prime matter, can also be separate from it, just like any other accident of a subject. Regents, as we will see, disagree with this: the explanation of this disagreement might lie in the different accounts of quantity as an accident or as a proprium of prime matter. So, prior to the qualification of this relation, which is the object of the next chapter, an as yet unqualified relation between matter and quantity may here be stated. According to the regents, natural compounds are quantified in so far as they are material, and the opposite holds too: material compounds are quantified. Consequently, forms acquire quantity as forms of material compounds. Considered alone, form is devoid of materiality and quantity: form is an indivisible and immaterial principle, and it can be regarded as material and quantified only when affected by the other principle of compounds. The union between matter and quantity is then stronger than the union between form and matter: this is evident because form and matter are independent principles, while quantity is a property of matter. Stevenson 1629 claims that the:

> species, quam forma tribuit materiae, adventitia est, et quasi extrinseca; quae cum ex se sit pars distincta a forma, ut potentia ab actu, habet per se speciem suam incompletam et invariabilem suamque unitatem specificam, quam non tollit diversitas specifica formarum quasi materialis et inadaequata, cum conveniant in una formali adaequata ratione sub qua referuntur ad potentiam materiae. [TP VII.3]

With respect to matter, form is something 'extrinsic' affecting it 'from outside' (*adventitia*). Matter itself already enjoys a proper specific unity, so form cannot give specific unity to matter. Furthermore, this specific unity is preserved through the specific diversity conferred by form. This diversity is somehow added to the existing specific identity of matter. Thus, the theory that:

> major igitur est unio inter quantitatem hanc et materiam, quam inter materiam et formam substantialem, saltem secundum quid. [King 1612, TP 3.V]

is grounded on the notion of a metaphysical act proper to prime matter and not dependent on form. The qualification '*secundum quid*' is intended by King to limit the validity of the statement to matter and form considered alone, that is, not while in a compound. Without the qualification, the unity of the compound would result in being accidental, posterior to, for example, the unity per se between prime matter and quantity. Matter and form can never exist one apart from the other: the really existing being is always the compound, not the two components alone. Yet, 'by consideration of the nature of matter', the union with quantity is logically prior to the 'extrinsic' union with form.

2.2 Prime matter: quantity and accidents

> Forma non est patibilis per se, sed quatenus in materia. Compositum patitur quidem; non tamen quatenus ex materia et forma constans, sed solum quatenus habet materiam. Nec sola forma, nec compositum, est subiectum cui inhaerent accidentia materialia quae de novo producuntur. Ergo in sola materia inhaerent. Materia ad recipienda accidentia non exigit formam, ut concausa receptionis passivae in eodemmet causae genere. [...] Forma accidentalis pro sui inhaerentia praesupponit formam substantialem, non tamen ei inhaeret. [Fairley 1615, TP XX]

Regents hold that there are accidents which inhere directly in matter, in virtue of which they subsequently inhere in the compound.[16] This theory should be understood in the light of claims that regents make concerning resolution and substantial form. In the process of natural corruption, if it is not true that all accidents inhere in the substantial form which gives actuality to the compound (this being the position of Stevenson 1629), then the problem arises of what the subject of these accidents is.

The two main solutions offered by regents are the following: 1) a minority holds that all accidents inhere in the respective substantial form of the compound: thus, the corruption of a compound is the dissolution of the relation between a form and its matter. Accidents cease to inhere in matter since there is no form by means of which they can inhere in it. This might be the solution given by Rankine 1627. A more widely accepted solution is 2) that some accidents inhere in matter immediately, without a substantial form, qua accidents directly flowing from quantity. So, the corruption of a compound does not entail immediately that the totality of accidents is corrupted, but only that the accidents directly

[16] See also, for example: Adamson 1600, TP IIII.2-3; Fairley 1615, TP XX; Mercer 1632, TP XII.6.

flowing from form are corrupt. The previous passage by Fairley opens and closes with a reference to heat (*calor*) and the way in which a compound can be said to receive heat:

> Si daretur calor separatus a materia nihil pateretur. [...] Ut materia possit calorem recipere satis est quaelibet forma specifica. [TP XX.1-10]

Heat without matter does not affect a compound, indeed it is not a physical phenomenon. Heat requires matter in order to affect a compound, but also any material form is enough to make matter receptive to heat. Thus, heat does not affect matter insofar as matter is informed by a form specifically apt to receive heat; on the contrary, matter informed by any form whatever is receptive to heat. The role played here by form is simply to give formal existence to matter (which cannot exist without form), not to make matter in any way receptive to heat in virtue of some specific formality. We can say then that a compound is heated or cooled only in so far as it is material.

Regents think that 'being hot' is a property of compounds immediately (because only compounds can 'be hot') but also that this property is grounded in matter, not in form. Matter in general provides the material cause of the process of heating; any material compound is potentially receptive to heating in the same way, because the underlying matter is the same. This is an important physical consequence of the identity of the material principle among all compounds.[17]

It seems clear that matter can be the subject of accidents. Regents call the accidents flowing from matter 'material accidents', distinct from 'formal accidents', due to form. In the categories, accidents flowing from quantity are 'material accidents', accidents flowing from quality are 'formal accidents': quantity depends on matter, quality on form. Granted that only form can provide physical actuality to matter, it follows that accidents inhering in matter receive from form physical actuality. Before information, these accidents are said to be 'interminate' (*interminata*), 'without a terminus'. Baron 1627 explains this point well:

> Cum omnis terminatio materiae, et quantitatis proveniat a forma, quantitas a materia profluens, ut talis, non alia esse potest quam interminata quae licet

[17] A question might be raised regarding the four elements: it is not possible to say that the matter of fire is receptive to heat in the same way as the matter of earth is. This is a general problem in the reception of Aristotle, and in the supposed unity of natural philosophy. In fact, commentators always pointed out the difficulty of reading the *Physics* in the light of the *De generatione et corruptione* and vice versa (G. Giardina, *La Chimica Fisica di Aristotele*, Rome, Aracne, 2008, chapter 1). The regents do not address this difficulty directly: I suppose that the answer might be that what has been said about matter and heat applies to physical substances as a mixture of the four elements, which are never found separate. Thus, prime matter would be the result of the mixture of the four elements.

> in rerum natura semper terminata existit, spectata tamen in essentia sua et quatenus a materia profluit, nullis terminis definita est, sed indifferens ad omnes. [...] Quantitatem igitur interminatam, materiae coaevam a Thomae sectatoribus, immerito explosam, nos cum Averroe Zabarella, et alijs magis nominis Philosophis, jure merito retinendam censemus. [TP III.6]

If all termini come from form, material accidents before information must be without termini; this does not mean (against Thomas), that all accidents come from form, or that all accidents are not actualised before information: in fact, these accidents qua interminate are rooted in the metaphysical act of prime matter and a compound is affected in such and such a way also because of these accidents. Among these accidents, extension is central, as will become clear in chapter 4. Regarding the relationship of quantity and form, Strang 1611 tells us that:

> formam materiatam necessario extensionem ac quantitatem requirere, eidemque continuitatem non nisi ex accidente competere. [TP IV.2]

A material form is the form of a compound: this form is necessarily quantified and extended by accident, not as form, but as form-in-matter.

2.3 Unity of the compound

Since compounds are the only natural beings which have existence in act, for this reason they are properly called 'substances'. Form and matter are 'incomplete' substances, because they exist only as principles of complete substances: their union, which makes up for their respective incompleteness, yields a complete substance. It is then clear that any discourse in natural philosophy has the substances as proper objects, and form and matter as objects only insofar as they are principles of these substances. In general physics (the branch of natural philosophy which deals with the principles of the natural world) form and matter are analysed separately one from the other not because they can exist in such a way, but only because the knowledge of components instructs us on the nature of the composition.

In reading graduation theses on general physics, the problem arises of what sort of unity is proper to natural substances: in the context of a metaphysics of essence, which ascribes an act to both form and matter, the essential unity of the compound is granted by the notion of ‘incomplete’ acts originating a ‘complete’ act. The stress on the components does not entail the priority of the components over the compound. Nonetheless, when it comes to analysing the properties of form and matter, we might find the philosophical justification for a weak dualism within the compound: both form and matter are subjects of properties, mutually dependent with respect to a compound, mutually independent with respect to the essence.

The influence of the philosophy of Aristotle is strong in the *Theses philosophicae*. Famously, Aristotle’s philosophy is centred on the concept of substance, as the first and ultimate being. A more thorough discussion of this matter will be possible after our analysis of movement. Regarding this first outline of the structure of substances, I believe that we can find two main tendencies among the regents, which are two sides of the same coin. First, only natural substances are complete substances, so proper activity and existences can only be predicated of them, not of their components. Secondly, regents sharpen the focus on form and matter as subjects of properties, in order to investigate the properties of the compounds in relation to their respective immediate substrata.

These two approaches are not exclusive, and they are often present within the same regent. This is why statements as the following:

> Forma et compositum non terminant diversas actiones; sed unam tantum, quae intrinsece terminatur ad formam, extrinsece ad totum compositum. [King 1624, TP VI.1]
>
> Nulla forma speciem, aut numerum dat materiae, sed toti composito. [Reid 1618, TP II.4]

should not be seen as contradictory. In fact, form and matter, as functional concepts, always refer to the compound and to one another, because it is only in a compound that form and matter become complete. As in King 1624, the action of form is the same as the action of the compound; not simpliciter though, because the action of form must refer intrinsically to form, and only extrinsically to the compound. How should this ‘extrinsically’ be understood? The stress on form does not endanger the unity of the compound; logically form is not the compound. In Reid 1618, in agreement with other regents quoted above, form does not specify matter, but does specify the compound of which matter is the material cause. Any activity of form on matter and of matter on form can only occur in the compound; nonetheless, the essences of its components entail an essential unity but not an essential identity between them.

Two distinct narratives are discernible here: some regents favour a stronger identity theory,[18] thus stressing the role of substance prior to its components: usually this view is obtained by reducing the activity of substances to form. An evident case is the acceptance of the Aristotelian doctrine that form is the nature of a substance (which I treat in part II, chapter 2); otherwise, other regents underline the equally important role of the two principles of substances,[19] as in the case of quantity extended per se, a significant contribution of the Scottish regents in response to the philosophical analysis of Transubstantiation (part I, chapter 4). One aspect does not prevail over the other, because the ultimate way in which forms and matter exist is as form and matter of a compound. The notion of *forma mistionis* can profitably be brought to bear here. At first sight the idea that matter and its accidents remain after the corruption of their substance could be interpreted as a strong statement in favour of the existence of matter independent of form. But it is not so, because a form proper to matter is still required in order to justify the ordered structure that we acknowledge in this portion of matter deprived of its substantial form.

A proponent of the first narrative is Rankine 1627. He does not talk of form of mixture, and seems to hold that there is only one substantial form within each compound, as we have seen. He also holds that matter is nature not secundum se, but as form itself [*ut eadem forma*]: two natures in the same compound are not possible [TP VI.4]. Yet, in thesis IV.4 Rankine touches on a much debated theory, once again of Scotistic origin:

> Licet igitur forma, compositum, et modus unionis, sint entitates realiter distinctae, non tamen requirunt distinctas actiones per quas producuntur, cum solum compositum habet esse per se.

This passage is very dense. The regent expounds his theory of the unity of natural substances: 1) form, compound and mode of union [*modus unionis*] are really distinct entities; 2) yet, they are produced by the same action: that means, the eduction of form from matter is identical with the production of the compound and of the mode of union; 3) because only the compound has existence per se, form cannot exist independent of matter. The interesting remark is the talk of 'mode of union'. The regents usually reject this notion, on the basis of the Aristotelian theory of substance, which does not accept a third

[18] For example, Robertson 1596, TP 10; Reid 1610, TP 2 and 1622, TP 6; King 1624, TP VI; Baron 1627, TP 3; Rankine 1627, TP VI (to be contrasted with IV); Wemys 1631, TP XIII.

[19] For example: King 1612, TP 3; Rankine 1627, TP IV (to be contrasted with VI); Murray 1628, TP II; Mercer 1632, TP XII; Leech 1634, TP IV.

entity of such a kind within the compound: in order to yield a unity per se of a compound, a substantial form and matter suffice.

Rankine seems to be close to the position of Suárez, who inscribes himself within the Scotistic tradition. Suárez writes that:

> distinguitur ergo materia a forma tamquam res a re. Et confirmatur nam compositio substantiae ex materia et forma est realis et physica [...] ergo ex duabus rebus. [*DM*, 13, IV, 5][20]

According to Suárez then, a third element is required in order to convey a unity: a 'mode of union' between form and matter, which are regarded as extrinsic principles. It is arguable that Suárez was influenced by the Augustinian tradition, to which Scotus belongs.[21] Even if Rankine's position is not accepted by many other regents, it is interesting to note that Rankine is still part of the Scotistic tradition: simply of a different one. I believe that this is further evidence for the influence of Scotus on the regents, and, more generally, on much of the Scholasticism of the seventeenth century.

One final remark helps us to qualify the theory of substance of the regents as a metaphysics of essence. According to King 1616:

> differentia individuans, etsi quidditas seu essentia non appelletur, cum non attingatur in definitionibus, nihil tamen impedit, quo minus sit pars essentialis individui. [TL V]

The individuating difference of a substance is part of the essence of the substance, even if it is not properly called 'essence' and it is not part of the definition. Only individuals exist, qua individual essences.

---

[20] Quoted in L. Prieto López, *Suárez, crocevia nella filosofia tra medioevo e modernità*, p. 15. The author claims that the account of the unity of the substance in Suárez anticipates and paves the way for the dualism of modern philosophy (in particular of Cartesian philosophy) because form and matter are different things and according to the Thomistic principle no unity per se is possible between two things in act. The interpretation of Prieto López is heavily influenced by Thomistic philosophy, and by the interpretation of modern philosophy as the historical moment of the breaking down of the unity of substances, of the forgetfulness of being and of the victory of phenomenalism.

[21] *Ivi*, pp. 12-13.

3. Conclusion

The majority of regents come close to attributing an ordered structure to matter without form, in virtue of the eduction of material forms, the metaphysical act of prime matter and the matter as subject of properties and accidents; yet, they could not bring themselves to adopt the theory of matter existing without form, since that would have required them to reject the Aristotelian metaphysics of substance. Dalrymple 1646 is an interesting case of breach of the Scholastic doctrine of prime matter as pure receptive potency. He reinterprets 'potency' as an active internal principle of change within matter, not as the receptiveness and indeterminateness of matter towards form. I argued that on this point Dalrymple has moved beyond traditional Scholasticism, while most of the other regents thought and taught in the Scholastic way in natural philosophy.

In this chapter I sought to expound the theories of the *Theses philosophicae* regarding the properties of prime matter. My focus has been on prime matter as a 'quasi-substance', namely as subject of properties in its own right, independent of form. The first part on resolution into prime matter has shown that regents accept the notion of form of mixture, which is the form proper to the body in the animate compound (including men) and the substantial form in inanimate compounds. In the former case, no resolution into prime matter occurs immediately in the corruption of the compound; in the latter, resolution occurs immediately.

The second part has dealt with the introduction of quantity as the key property of prime matter. In virtue of quantity, prime matter is the subject of properties which flow from quantity: as we will see in the next chapter, these properties include extension per se in place and divisibility. The focus on form and matter in the theses prompted the question of the unity of the natural compound: regents seek to preserve the essential unity of the compound by overlooking the Scotistic talk of mode of union and *haecceitas*, even if their metaphysics of essence brings them close to these notions.

De Transubstantiatione

1. Preliminary remarks

One of the most noteworthy features of the *Theses philosophicae* is the unanimous rejection of the Catholic doctrine of Transubstantiation. What the regents have to say on this matter is noteworthy historically and theologically, because it is a sign of the definitive fracture within the Christian world at the beginning of the modern era. It is also philosophically noteworthy, for, while rejecting a doctrine in itself theological, the regents not only employ philosophical tools, but also expound philosophical conclusions whose importance is paramount in order to understand the 'Reformed' character of the Scholasticism of the theses and also to shed some light on the relationship between Scholasticism in the early seventeenth century and modern philosophy.

The doctrine of Transubstantiation is a historical product of Christian theology concerning the interpretation of the evangelical episode of the Last Supper. In the three synoptic Gospels, Jesus, at the offering of the cup and the breaking of the bread among his apostles utters these words: *"This is my body which is given for you; do this in remembrance of me.* [...] *This cup is the new covenant in my blood, which is shed for you"* (Luke 22: 19-20, King James Bible). As far back as Thomas Aquinas, these words are usually interpreted by the Church of Rome to mean that bread and wine really became the body and blood of Jesus. In the words of Thomas:

> hinc autem manifestum est quod in conversione praedicta panis in corpus Christi non est aliquod subiectum commune permanens post conversionem: cum transmutatio fiat secundum primum subiectum, quod est individuationis principium. Necesse est tamen aliquid remanere, ut verum sit quod dicitur, *hoc est corpus meum*, quae quidem verba sunt huius conversionis significativa et factiva. Et quia substantia panis non manet, nec aliqua prior materia, ut ostensum est: necesse est dicere quod maneat id quod est praeter substantiam panis. Huiusmodi autem est accidens

> panis. Remanent igitur accidentia panis, etiam post conversionem praedictam. [*CG*, I, 4, 63]

Leaving the philosophical considerations aside for the moment, it is important to notice that Thomas underlines the importance of the words *hoc est corpus meum* as univocally meaning that the bread turns into the substance of the body of Jesus - and the same for the wine, which turns into the blood of Jesus. The doctrine of Transubstantiation entails that during the mass the officiant calls for God's miracle of changing what originally are bread and wine into the real body and blood of Jesus: this change does not affect the external appearance of bread and wine, which retain some of their original characteristics, such as flavour and colour.

Scholastic philosophers always faced the challenge of accounting for this miracle in a way which was intelligible in terms of philosophical rationality, without questioning the truth of the dogma based on the authority of the Gospel. This is a case of the broader debate revolving around the relationship between philosophy and theology: Scholastics hardly abandoned the Thomistic slogan of philosophy as the *'ancilla theologiae'*, 'maid servant' of theology.[1] More precisely, they held that any true proposition in philosophy can be true only if in agreement (or not in contradiction) with an authoritative proposition in theology, while the opposite is not required. This way, philosophical propositions can be divided into propositions 1) in open contradiction with theology [for instance, 'the world is eternal']; 2) in agreement with theology ['the world is created']; and finally 3) neutral with respect to theology ['world is composed of matter and form']. Propositions of type 1 are not acceptable in Scholastic philosophy: much of the opposition to Aristotle from the twelfth century onwards highlighted those of his doctrines that contradict the Bible. Propositions of type 2 are acceptable and philosophically fruitful, because they show the inner harmony between natural reason and revelation. Propositions of type 3 are acceptable and can be fruitful: the example of the universal structure of matter and form is an Aristotelian cornerstone of many Scholastic systems.[2]

The doctrine of Transubstantiation is surely a philosophical product because we do not find it in the scriptures in a philosophical form (namely, shaped in the form of Aristotelian philosophy). It took form first in the Eastern Roman Empire in a context of Platonism and Aristotelianism, and was then fully accepted and strengthened by Scholastics in the Middle Ages. Thomas's formulation enjoyed great success also because of the official

[1] Formula which is repeated in R. Baron's *Philosophiae theologiae ancillans*, St Andrews 1621.

[2] The relation between 'natural light' and 'light of the faith' is treated by Baron in exercitatio III of the *Philosophiae theologiae ancillans, passim,* in particular art. VII and arts. XXIV-XXVI.

endorsement by the Council of Trent (1543-1568), after which it became the official formulation of the Catholic Church, accepted until now. For this reason, the doctrine of Transubstantiation is also a historical product, which awaited only the decision of a church council to be definitive in words and spirit once and for all. Before that, many concurrent versions of the explanation of the dogma were available, all of them equivally valid insofar as they all referred back to the letter of the Gospel; yet all different, according to the individual philosopher who formulated them. For instance, Thomas's and Scotus's accounts of Transubstantiation are equivally valid theologically because both admit the real presence of the body and blood of Jesus, yet they are not the same account because they reach the same conclusion in different ways and within different philosophical systems.[3]

It is not possible to prove that regents had Thomas's account in mind when writing against the Catholic dogma of Transubstantiation: what appears from the texts is that their main (yet not unique) opponent is Francisco Suárez, who held a position similar to Thomas, and who was arguably the most important Catholic voice at the time of the regents. The position they address will be clear from their own criticisms, but a few preliminary philosophical remarks will be useful. In the passage quoted above we find *in nuce* all the most important features of the philosophical account of Transubstantiation: 1) the conversion takes place at the level of substance, so it is a total conversion of one substance into another, leaving no room for the coexistence of two substances (i.e. bread and body) in the host; 2) what remains after the conversion are the accidents of bread and wine, as Thomas explains it, *'ut verum sit quod dicitur'*: these words can be taken to refer to both the conversion and the preservation of accidents. This text explains the nature of the conversion and hints at the most debated difficulty about Transubstantiation, the preservation of accidents.

According to the Medieval Scholastics, sense-data (*sensibilia*) when apprehended by their proper sense do not deceive us: in a formula, *sensus circa propria sensibilia non decipitur*. Experience testifies so firmly to the presence of the original characteristics of both bread and wine after the conversion that any account of Transubstantiation must include a justification of this preservation. The first step in this direction is taken when

---

[3] *"The doctrine of transubstantiation, first declared orthodoxy at Lateran IV, might be said to be fully explicated only among the theologians at the Council of Trent"*, L. P. Wandel, *The Eucharist in the Reformation*, Cambridge, Cambridge University Press, 2006, p. 219. The theologians of the Council of Trent did not accept Luther's formulation of the distinction between the 'real presence' and 'transubstantiation' in the host. John Knox's position (*ivi*, 184-192), very influential in Scotland, accepts the 'real presence' but does not accept transubstantiation. It is noteworthy that in Knox and in the regents the same question of the sacrament of the host is answered in two different ways: one theological, the other philosophical. Knox seems to be content with a formulation which could not satisfy a Scholastic philosopher.

Thomas denies that the matter of bread (or of wine) could remain through conversion. He reasons as follows: matter could be the substance of these remaining accidents, preservation of matter would then explain the preservation of accidents. Yet, matter only exists in virtue of form, because the substantial form is the act of matter, which is pure potency. If matter remained, then its form would remain as well, since matter alone is nothing. In which case the conversion, far from being explained, would have been rejected. In *CG*, I, 4, 65, Thomas addresses the issue of the accidents:

> nec est impossibile quod accidens virtute divina subsistere possit sine subiecto. Idem enim est iudicandum de productione rerum, et conservatione earum in esse. Divina autem virtus potest producere effectus quarumcumque causarum secundarum sine ipsis causis secundis: sicut potuit formare hominem sine semine, et sanare febrem sine operatione naturae. Quod accidit propter infinitatem virtutis eius, et quia omnibus causis secundis largitur virtutem agendi. Unde et effectus causarum secundarum conservare potest in esse sine causis secundis. Et hoc modo in hoc sacramento accidens conservat in esse, sublata substantia quae ipsum conservabat.

In the normal course of nature, no accidents can be without their substance; in the miracle of Transubstantiation 'it is not impossible' that God by *potentia absoluta* maintains these accidents once their substance is destroyed. God cannot create mutually contradictory effects but he can produce an effect without its (secondary) cause. This general principle implies more than Thomas spells out in this passage: the reference is to the theory of the dependence of all creatures on God as metaphysical primary cause of all things. This bond cannot be broken, while the bond between created things (viz. between a substance and its accidents, or between a cause and its effect) can be broken, even if only by God.

Transubstantiation is therefore a substantial conversion of one substance into another, where the accidents of the former substance are preserved, as experience shows and philosophy explains.

Quantity is an accident of matter. In the Scholastic sources of the sixteenth and seventeenth centuries, the role of quantity in Transubstantiation is evident. Starting from the inclusion of quantity in the number of accidents which can exist apart from their substance, in the seventeenth century it was common doctrine that quantity acts somehow as a 'quasi-substance' in which the other accidents continue to inhere once their substance (the compound of form and matter) is dissolved. Suárez holds that:

> in mysterio Eucharistiae Deus separavit quantitatem a substantiis panis et vini, conservans illam, et has convertens in corpus et sanguinem suum; id autem fieri non potuisset, nisi quantitas ex natura rei distingueretur a substantia. Neque sufficere potuisset distinctio modalis, quia substantia non potest esse modus quantitatis. [*DM*, 40, II]

Quantity as an accident must be different from its substance (no substance is identical with its accidents); more precisely, by the truth of Transubstantiation, quantity must be different from substance *ex natura rei*, as thing from thing, because not even God can make it that a mode and its substance are separate.[4] And it is true that in order to make Transubstantiation intelligible and not only accepted by faith, quantity is separate from its substance.

The connection between the theory of prime matter and Transubstantiation is clear in virtue of the role played by quantity: quantity is an accident of matter, thus any relation between quantity and matter influences the possible account for Transubstantiation. Different relations imply different accounts. And clearly the regents and Suárez did not agree on this matter. The philosophical relevance of this seventeenth century debate is not limited to this question but it extends to other key Scholastic doctrines, such as the notions of accident, substance and place. I take the regents as intentionally distancing themselves from what they considered to be "ad hoc doctrines" that had been devised for the purpose of justifying a theological dogma, against what the regents call 'good philosophy'.

The entirety of the debate on Transubstantiation cannot be dealt with in this context. I shall follow the *Theses philosophicae* in order to expound the criticisms that the regents put forward but also to present the theories they oppose, when the regents themselves fail to do so. This account may not be inclusive of all the qualifications of the debate but it will cast light on the principal moves in the debate. The first notion to explain is that of accident: what is included in the definition of accident, whether its definition includes inherence in a substance and what sort of inherence it includes. I shall compare the definitions of accidents in standard seventeenth-century Scholastic texts with the definition that regents provide. I

[4] *"Si alterum extremorum ex illis duobus tale est, ut per potentiam Dei absolutam, non possit sine alio conservari, magnum argumentum est, illud essentialiter tantum esse modum quendam, et non veram entitatem; quia si esset vera entitas non posset habere tam intrinsecam dependentiam ab alia entitate, ut non possit Deus illam supplere sua infinita potentia: ergo solum potest id provenire ex eo, quod illud extremum in sua intrinseca essentia non est entitas, sed tantum modus"*, Suárez, *DM*, 7, II, 7-8. A mode of a substance cannot exist without its substance and not even God can bring about that it does. I understand the expression *'ex natura rei'*, which in *DM*, 1, VII, 13-20 is employed to describe the modal distinction, to mean the real distinction, in order to make sense of the following: 'neque sufficere potuisset distinctio modalis.' In conclusion, according to Suárez, quantity is not a mode of a substance because God can bring about that it exists without its substance, therefore they are really distinct (*DM*, 40, II, 1).

will then move to the analysis of a particular accident, quantity, which, as we have seen, is the category in virtue of which other accidents can inhere in matter, such as the colour and the flavour of the bread of the host. In this analysis, the notions of extension and place will become central, because the regents disagree with Catholic Scholastics on both points. I shall argue that the regents develop theories of the relationship between matter and quantity and between quantity and extension in place which are coherent with their Reformed reading of the words *hoc est corpus meum*.

2. Separability of the accidents

The dogma of Transubstantiation famously influenced the development of the philosophy of Descartes, who took pain to ensure that his system was compatible with the teaching of the church in his replies to Arnauld. Scottish regents too dedicate many lines to analysing the philosophy Catholic Scholastics used to make sense of their faith. It is primarily a matter of faith: regents belonged to the Reformed Church of Scotland, which rejected the dogma of Transubstantiation, and offered a different reading of the passages in the Gospel that Catholics read as a *verbatim* proof of such miracle. The Scottish position was not accepted by all Reformed churches, but Scottish reformers developed their national church from Calvinist elements and offered a symbolic reading of the host.[5]

Both the Catholic and the Scottish Reformed positions are inevitably influenced by a prior and pre-philosophical acceptance of a specific faith and the role of philosophy is to provide clarification of and perhaps also support for the faith. At bottom, regents and Catholic Scholastics go down the same path, and if scholars (often looking at Scholasticism from the standpoint of modern philosophy) criticised the Catholic justification of Transubstantiation as "ad hoc" or theologically motivated, I do not see why the same cannot be said about the regents. Yet, as we will see, the regents, in their criticism of the Catholic position, develop a theory which anticipates modern philosophy. The question is not whether these theories are theologically motivated or not: because all of them are; not even whether this is a licit move in philosophy. The question is rather how fruitful this relationship between theology and philosophy has been. I believe that the regents actively worked on their philosophy inspired by their faith.

[5] For example, Robert Bruce, sermon *The Lord's Supper in Particular 1: 3. The things contained in the Sacrament*, in R. Bruce, *The Mystery of the Lord's Supper*, edited by T. F. Torrance, Edinburgh, Rutherford House, 2005, pp. 70-90.

The accusation that the doctrine was being sustained by ad hoc philosophical principles is not completely off-target. Consider, for example, Suárez writing about the separability of quantity from matter, therefore about their real distinction:

> *Approbatur sententia reipsa distinguens quantitatem a substantia.* Atque haec sententia est omnino tenenda; quamquam enim non possit ratione naturali sufficienter demonstrari, tamen ex principiis Theologiae convincitur esse vera, maxime propter mysterium Eucharistiae. [*DM*, 40, II, 8]

Suárez's opinion is that the separability of quantity from substance cannot be grounded on pure natural reason but is in need of a theological justification, which nonetheless opens up the way for philosophy in its attempt to justify it. Suárez cannot offer any other example of quantity deprived of its own substance, nor of substance deprived of its own quantity: all examples refer to Transubstantiation and related philosophical corollaries (for example, the presence in the host of the body of Christ without its actual dimensions).

2.1 Definition of accident in a standard Catholic theory

As mentioned earlier, quantity is an accident: when considered qua accident it must fall under the definition of accident, traditionally established by Porphyry in *Isagoges* 12, 23-25: *"Accident is what can be present or absent in a subject"*, without implying the destruction of the subject.[6] This definition is found in an introduction to Aristotle's *Categories* and it inevitably reflects the vast debate over the real nature of this treatise. Whether the *Categories* are originally a logical or a metaphysical work, or both, Scholastics used to interpret it as an ontological work that shows how we classify things and how things really are, and establishes a harmony between knowing and being. Porphyry's definition is generally accepted by Catholic Scholastics. Eustachius a Sancto Paulo is clear about this point. In the logical part of his *Summa philosophica quadripartita* (I, II, V) Eustachius makes this definition his own from a logical point of view, without moving any further. In the metaphysical part instead, he takes on the problem of the separability of accidents in metaphysical terms, arguing that:

[6] '[Σ]υμβηβεκός ἐστιν ὃ ἐνδέχεται τῷ αὐτῷ ὑπάρχειν ἢ μὴ ὑπάρχειν.' My translation.

> inhaerentiam quidem aptitudinalem in formali ratione accidentis contineri; verum inhaerentiam actualem saltem ex natura rei ab accidentis natura seu essentia esse diversam. [...] Quod autem inhaerentia proprie dicta, quae est actualis, diversa sit ab accidentis essentia, ex eo liquet, quod ratio accidentis posita sit in eo quod sit forma subjecti completi seu totius compositi actu existentis. [...] Ex quo intelligis inhaerentiam non esse rationem formalem accidentis, sed modum existendi ipsius naturalem. [*SPhQ*, pars IV, Tractatus de principiis entis, II, VIII]

In logic, the difficulty concerning separability is overcome by distinguishing between *proprium* and *accidens* and ultimately, when it comes to inseparable accidents (such as the whiteness in a swan, following Eustachius's example), Eustachius claims that they are separable when we consider the subject as species, not as an individual.[7] In metaphysics the appeal to species is not available, because the separability of accidents concerns one single individual in its own structure. The philosophical tool by which this solution is acquired is the Scholastic notion of *inhaerentia*, divided into *actualis* (actual) and *aptitudinalis* (aptitudinal). An accident inheres in its substance in the vast majority of cases: more precisely, in all *physical* cases. Yet, actual inherence cannot be mistaken for inherence per se, neither for the formal reason of the accident, because in no way can the accident be defined by its inherence in a particular substance. Eustachius holds that "the reason of the accident is that it is the form of the complete subject or of the whole compound existing in act." The actual inherence is only a mode of existence natural to the accident, not its definition. It is a mistake to take a mode for the definition, since it goes against the logical principle *e dicto secundum quid ad dictum simpliciter non valet illatio*. The more correct notion (more correct because coherent with all the possible instances of the existence of accidents) is that of 'aptitudinal' inherence, which refers to actual inherence as not included in the essence of the accident. While actual inherence is different from the essence of the accident as thing from thing, aptitudinal inherence cannot be separated from the accident; it is included in the essence of the accident:

> sicut enim fieri nequit ut accidens non sit aptum inhaerere, sic etiam evenire potest ut interdum actu non inhaereat, licet nihil, quoad ad ejus essentiam attinet, immutetur. [*ibidem*]

[7] Scotus raises the question of the identity between inseparable accident and proprium in the quaestio 32 of his *Quaestiones in librum Porphyrii Isagoge*: 'Utrum proprium sit distinctum universale ab accidente.' Scotus's answer is that proprium and accident are two distinct universals because they do not have the same definition: the proprium cannot *adesse* and *abesse*.

So, accidents are separable from their substance because it is not actual inherence but only aptitudinal inherence that is part of their essence. Therefore, Scholastics developed a theory of accidents compatible with the non-natural occurrence of an accident not inhering in its substance, as in the case of Transubstantiation.

Eustachius is not directly mentioned in the theses, while Suárez is. Suárez will play a major role later on, while Eustachius's exhaustive style proves very useful for clarifying the starting point, and made the fortune of his main work which I am quoting, the *Summa philosophica quadripartita*. Due to the nature of the *Theses philosophicae*, in this case as in many others the regents do not dedicate much room to the exposition of theories other than theirs, and this work is left for the reader. I consider Eustachius a useful source for an exposition of what can be taken as the general framework of a Catholic account of Transubstantiation.

2.2 Definition of accident in the *Theses philosophicae*

Regents usually treat the notion and definition of accident in the *Theses logicae*, the section dedicated to logic, in accordance with the origin of the debate, Porphyry's introduction to the *Categories*. They never treat it in metaphysics, and when it comes to physical theses all the work is done on the basis of what has been previously said in the logical theses; in the theses we do not see the shift in analysis from logic to metaphysics as in Eustachius.

The definition of accident is usually expressed in traditional terms and the definition by Porphyry is never rejected. In principle, regents agree that the characteristic of accidents is that it can be or not be in a substance, without changing the definition of the substance itself. Where they stand apart from Catholic Scholastics is not with respect to the general notion of separability of accidents, but the separate existence of accidents. By 'separate' regents do not mean 'an accident existing in a substance other than its original substance', or more generally, 'an accident without its own substance'. They seem to shape the problem around the very idea of a separate accident with 'separate' meaning 'without *any* substance'.

The words of Leech 1638[8] are very clear: *"Accidens existere posse se solo extra subjectum, manifeste implicat* [contradictionem] / *Ad accidentis solidiorem realitatem stabiliendam actualem in subjecto inhaerentiam adscribimus"* [TL 26-27]. In this passage the regent is expounding two key features of his notion of accident that many other regents agree on: 1) an accident cannot exist *se solo*, by its own powers, outside a subject; 2) an accident has actual inherence in a subject, the same actual inherence denied to it by Eustachius. We find in Seton 1630[9] a similar theory, appealing to the authority of Averroes: *"Inhaerentiam actualem, quam ab aptitudinalis nihil differre putant, de accidentis (quantitatis nimirum aut qualitatis) esse essentia, Averroes ejusque sequaces affirmare non verentur"* [TL 17].

In particular, I wish to focus on two longer passages, the first one by Stevenson 1629:

> Ad realem omnis accidentis existentiam, requiritur actualis inhaerentia in subjecto, nec sufficit aptitudinalis [...] / Licet multa dentur accidentia separabilia, sine quibus subiectum potest existere, nullum tamen datur separabile, quod sine subiecto existit, aut existere potest. / Adeo, ut illud, accidentis esse est inesse, de actuali inhaerentia, et reali existentia praecipue intelligatur. [TL XVI]

and the second one by Baron 1627:

> Essentia rei non recipit magis et minus, sed omnino in indivisibili consistit. Ergo inhaerentia actualis non est de essentia accidentis: haec enim admittit intensionem et remissionem. / Cum igitur accidentis esse sit inesse, inhaerentia aptitudinalis erit propria ratio et essentia accidentis. / Ut subsistentia se habet ad Substantiam, ita inhaerentia actualis ad Accidens, h. e. non est ipsa ejus existentia, et longe minus

[8] David Leech (ca. 1600- ca. 1657/64), minister of the Church of Scotland and regent. MA at King's College in 1624, under John Forbes. Leech's name is listed in the Latin form 'David Leochaeus' in Forbes's *Theses philosophicae*. According to the *DNB* he was appointed regent in 1628, while the *FAM* reports 1627. We have Leech's graduation theses for the years 1633, 1634, 1635, 1636, 1637, the same year when he published the academic oration *Philosophia Illachrymans*, and 1638 (the *DNB* does not list the 1638 theses, and reports the wrong title for the 1637 theses). He initially refused to sign the covenant and his later conversion to it was not fully convincing. After leaving university due to his initial rejection of the National Covenant, Leech lived between the army and the church. Created DD by Aberdeen University in 1653, he never returned to Scotland, and died after 1657, when he is last recorded in London. *DNB*.

[9] John Seton, MA in 1616, probably at Marischal College, Aberdeen, as a 'Iohannes Setonus' is mentioned in the list of graduants in Aedie 1616. Seton took the position of James Sibbald as Professor of Natural Philosophy in 1626, at Marischal College. As for his predecessor, we have a list of graduation theses written by Seton which do not follow the four-year curriculum: 1627, 1630, 1631, 1634, 1637 and 1638. Seton graduated classes which studied under different regents as well. *FAM*, p. 34: the graduation theses of 1638 are not included in the list in *FAM*.

> essentia, sed tantum existendi modus. / Non minus impossibile est accidens existere extra omne subjectum inhaesionis, quam substantiam non subsistere, sed alteri inhaerere. / Aptitudinalis inhaerentia accidentis, non vel per ipsum Dei potentiam absolutam, separari potest ab ejus inhaerentia actuali; quoniam hujusmodi separatio implicat contradictionem. / Nullum igitur praebet patrocinium absurdo Pontificiorum commento Transubstantiationis, et accidentis existentiae extra omnem substantiam. [TL X-XI]

Read alongside the two points mentioned before concerning accidents always being in a subject and concerning the ascription of actual inherence to accidents, these two passages yield important insights in the regents' position. Stevenson is intentionally using the word '*separabilis*' in two different senses, one logical, the other metaphysical. He claims that even if there are separable accidents according to the definition of accident, yet there is no separate accident in the metaphysical sense, that is, an accident which exists without a subject, and more generally an accident which could exist without a subject. Stevenson is even clearer when saying that '*accidentis esse est inesse*', de facto eliminating from the definition of accident the reference to 'adesse et *abesse*'. In a metaphysical sense, accidents cannot exist without a subject because their being is defined as 'being-in-something' to which they are related by actual inherence. The relation between an accident and its inherence in its own subject is one of identity.

Baron shows his knowledge of contemporary Scholastic texts by hinting at Eustachius's passage at length, until just before the definitive reference to Transubstantiation. Baron bases his idea that actual inherence does not belong to the essence of accidents on the fact that essences are immune from intension and remission, while actual inherence is not:[10] what Baron is saying is that the actual inherence of an accident can undergo degrees of change which cannot be included in an essence - by definition immutable. We find again the expression '*accidentis esse sit inesse*', which is typical in the *Theses philosophicae*. Despite the similarity in words and the agreement on aptitudinal inherence as the reason of accidents, Baron's stress on the *inesse* of accidents distances him from Eustachius.

The relation between inherence and accident is explained by an analogy of proportion presented as follows:

[10] Intension and remission (*intensio* and *remissio*) are the addition and subtraction degree by degree (*gradus ad gradum*), which imply a more and a less (*magis* and *minus*). They only occur in the category of quality, therefore they are qualitative 'more' and 'less'. In the category of quantity addition and

Subsistentia : Substantia = Inhaerentia actualis : Accidens

whose meaning is that actual inherence belongs to an accident just as subsistence does to a substance, that is, as a mode of existing, not as part of the essence. A substance does not entail existence in its essence, because it is a created and finite being. Existence is ultimately something which happens to a substance, not something a substance does essentially. What the substance does is be both in logical non-contradiction within itself (since a contradictory being cannot exist) and also in metaphysical subjective potency to being (the openness to existence of a substance before its coming into being). In the same way, actual inherence happens to an accident, but it is not something which the accident either is or does.[11]

In the second part of the passage, the differences with Eustachius become even more remarkable, and unbridgeable. Baron holds two theories Eustachius cannot agree with: 1) the separate existence of an accident is paralleled with the attribution to a substance of modes of existing which are per se a negation of the very definition of substance: both not subsisting and also subsisting in something else. This would turn a substance into an accident: it is a categorial mistake, as is a separate accident, which would become a substance. 2) The inseparability of aptitudinal inherence from actual inherence. These two inherences are not equivalent, because actual inherence is not part of the essence, while aptitudinal inherence is. Yet, they cannot be separated, the former being a mode of the second, as their separation implies a contradiction. Baron refers to God's absolute power, which is unable to perform the separation, while it was enough in Eustachius to ground the separate existence of accidents from their substances. I do not think that here the notion of absolute power is being questioned by Baron; the difference lies in the sort of task do-able by the exercise of God's absolute power: separating a mode from its substance is beyond God's powers, as also Suárez claims. The main point made here by Baron is the inseparability of actual inherence from aptitudinal inherence, which sets him apart from the philosophy structured with a view to justifying Transubstantiation.

I also believe that the analogy proposed by Baron is best explained by reference to the metaphysics of essence. In fact, just as subsistence flows from the essence of a substance, actual inherence flows from the essence or reason of an accident. Subtracting subsistence from a substance is as contradictory as subtracting actual inherence from an accident.

---

subtraction are called augmentation and diminution. In this context, intension and remission are not predicable of essences because they are unchangeable, if absolutely considered.

[11] Young 1613, TL 7.II: *"nullum accidens inseparabile subjecto suo necessarium est."* The actual inherence understood as part of the definition of the accident does not imply that an accident inheres in a substance necessarily: an accident is, by definition, accidental to whatever substance it inheres in.

All the four texts mentioned make use of a specific interpretation of the word 'separate' which is taken to mean, as I said, 'without *any* substance'. It is clear that this meaning is what regents see in the words of Catholic Scholastics, and this is the point that they reject. The regents' rejection both of Transubstantiation and of this notion of separability of accidents are not exclusively based on this strong interpretation of 'separate', which is likely to be rejected by Catholics too, since it seems to imply the idea of accidents really existing per se as substances. What regents reject is, more precisely, the process by which accidents are separated from their own substance and sustained without it by God's power: for the Catholics, this is the only way to account for Transubstantiation (in a sense, a breach in the normal course of nature). For the regents, this is an illicit move that contradicts the definition of accident. As Stevenson writes, *"accidens* ex Porph. *semper existit in subjecto, et* ex Arist. *non potest seursum existere ab eo in quo est"* [*ibidem*].

In conclusion, regents seem to include in the definition of accident the notion of the existence of accidents in their *own* natural substance, a notion which per se is not included in a traditional Scholastic definition. In fact, even if the reason of accidents prescinds from existence, their nature absolutely considered implies that they can only exist in a substance.[12] What regents do is to stress this characteristic and extend it to the reason of any individual accident. As we have seen, this has a dramatic effect on the concept of Transubstantiation, an effect which the analysis of quantity clarifies even more.

3. Quantity: its role in Transubstantiation and its relation to extension

Quantity in relation to prime matter has been treated already in part I, chapter 3. The conclusions reached there can be summarised in two key points: 1) quantity is a primary attribute of matter, which matter has independently of form; and 2) prime matter can be the subject of accidents in virtue of quantity.

These conclusions can be expanded by saying that quantity is essentially extended, that means that it has 'parts beyond parts' (*partes extra partes*): any *quantum* must be divisible into different parts. On this very general basis shared by all Scholastics differences are then developed by individual philosophers. The regents' debate over Transubstantiation starts from and expands the theory of the relation between matter and quantity.

[12] Thomas Aquinas, *Scriptum super sententiis*, I, d. 26, q. 2, a. 1, corp., quoted by M. Henninger, *Relations: medieval theories 1250-1325*, Oxford, Clarendon, 1989, p. 16. Thomas is referring to the 'absolute' accidents, quality and quantity.

As already remarked, what makes Transubstantiation special in Scholasticism is its having the nature of a breach in the natural course of events, a breach which has to be accounted for within the theory of natural substances. In Catholic Scholastic natural philosophy, the miracle of Transubstantiation cannot be left unaccounted for: the theory of natural substances seeks to explain the theological evidence of the separability of the accidents of bread and wine, even if Catholic Scholastics agree that Transubstantiation is not a natural event.

When it comes to quantity, one more question becomes central: quantity is what matter-related accidents inhere in, such as extension: thus, what does it mean that quantity is by essence spatially extended? The solutions that regents give to this question set them definitively apart from their contemporary Catholic colleagues.

3.1 Traditional views on quantity and extension

While the ten categories are usually divided into substance and nine remaining 'accidents', modern Scholastics further distinguished the nine accidents into quality and quantity as primary accidents and the remaining seven categories. The distinguished role of quality and quantity has been acknowledged since the thirteenth century, as we find it in Thomas and Scotus. Scotus calls them 'absolute accidents', introducing a terminology accepted up to the time of the regents, for example by Eustachius. Specifically, Scotus, while defending the notion of Transubstantiation, claims that absolute accidents can exist without a substance because they are not identical with their relation with their substance. It is then a case of real relation in which it is not contradictory that the foundations of the relation (viz. substance and absolute accidents) can exist without the existence of the relation.[13] Scotus's contributions in defence of Transubstantiation will be useful to us later on while we seek to clarify the theory of the regents.

Let us accept that quantity, as an absolute accident, enjoys the condition of being the subject of inherence of other accidents, namely those depending on matter. In the words of the apocryphal Thomistic text *Summa logicae*: *"quantitas autem licet sit fundamentum aliorum accidentium, tamen sequitur materiam"* [4, 5]. These accidents following from matter include the category of place, important for further aspects of Transubstantiation.

With regard to quantity, the Coimbrans affirmed that:

---

[13] This is an application of the principle of separability (two things are really distinct if it is not contradictory that one exists without the other) which Scotus definitively linked to the real distinction. M. Henninger, *Relations*, pp. 71-74.

> essentialem ac propriam quantitatis rationem consistere in extensionem partium, hoc est, ut quantitas ipsa in ordine ad se habet partem unam extra aliam [...] ita effectus formalis quantitatis est extendere partes materiae easque in toto ipso inter se ordinare ac distinguere. [*In Phys.*, 4, 5, 4, 2]

This general point is not questioned by the regents, who however raise many doubts with respect to further qualifications of quantity and extension, particularly by Eustachius and Suárez:

> Verum cum duplex esse possit extensio rei quantae: altera velut externa et sensibus perspecta, nempe extensio partium in ordine ad locum, altera vero interna, a sensibus plane remota, nempe extensio earumdem partium in ordine ad se, gravis hic difficultas oritur, quaenam extensio sit essentialis et intima ratio quantitatis. [...] Repugnet enim aliquid esse sine eo quod ad ejus essentiam pertinet;[14] quare necesse est rationem quantitatis in alio positam esse, nempe in extensione partium in ordine ad se. Et certe natura prius est partes rei quantae extensas esse simpliciter, seu in ordine ad se, quam in ordine ad locum, cum locus sit quid extrinsecum rei quantae. [*SPhQ*, I, III, II, I]

> Secunda ratio principalis ex mysterio sumpta est, quia sub speciebus consecratis est corpus Christi Domini cum sua naturali quantitate, et tamen non habet extensionem partium suarum in ordine ad locum, ut ex fide constat; ergo actualis extensio partium substantiae in ordine ad locum non est ipsa quantitas substantiae. [*DM*, 40, II, 14]

To avoid the evident problem of the body of Christ converted into a host with a much smaller extension in space, Eustachius and Suárez (among others, of course) develop the distinction between 'internal' and 'external' extension. The former is the type of extension that a thing has *in ordine ad se*, within itself; the latter is the type of extension we usually experience, *in ordine ad locum*, extension extended in place. Eustachius and Suárez agree that, in order to save the miracle of Transubstantiation, we must include in the essence of quantity only the internal extension, despite the fact that Transubstantiation could be the only occurrence where this distinction between internal and external extension actually

[14] The implicit reference is to Christ's body in the Eucharist without extension in ordine ad locum.

carries weight. Eustachius also reminds us that we can not include in an essence something extrinsic: in this case, place with respect to quantity.[15] The relation between place and quantity is between really distinct things. It is not contradictory that quantity is without place, nor therefore that the extended body of Christ is deprived of a fixed relationship with its extension in place, while retaining extension *in ordine ad se*.

3.2 Regents on quantity and extension

> Essentia et formalis ratio quantitatis in extensione partium consistit, seu in eo quod est habere partem extra partem secundum extensionem et molem. / Quod quantitas sit loco extensa [...] quodque sit impenetrabilis et hujusmodi, ei tantum conveniunt in ordine ad extrinsecum nempe locum. / Distinguenda igitur erit essentialis extensio partium quantitatis inter se, qua distinctam obtinent magnitudinem et molem, ab hac extensione in ordine ad locum, cum sine hac prior servari possit. / Et nihilominus substantia a quantitate separata esto a se, et ex se partes entitativas habeat, partes tamen extensionis et molis non haberet. / Quare cecutiunt ad lucem veritatis, qui asserunt separata quantitate a substantia corporea eam in eadem dispositionem permansuram [...] cum substantia corporea quantitate spoliata ad modum indivisibilem ratione loci reducatur, ita ut nullum prorsus locum occupet. [King 1612, TL 11]

> Quod itaque quantitas primo substantiae tribuit, non est extensio partium entitatis, sed molis, quae ex propria natura loci sunt occupativae. / Ideo essentialis ratio quantitatis ponitur in hac extensione partium molis. [Reid 1622, TL XIV.3-4]

> Quantitatem materiae inseparabili nexu cohaerere. [Baron 1627, TP III]

> Nec aptitudinalis extensio in loco, ut somniant Metusiastae, nec actualis, est essentia quantitatis, sed ponitur in extensione suarum partium, et partium substantiae, inter se et in toto. / Inconsiderate distinguunt quantitatem in internam et externam, prout partes in entitativas et quantitativas, qui quantum illocaliter esse volunt. [Mercer 1632, TL IX.4-5]

---

[15] I believe that this remark is similar to the exclusion of actual inherence from the essence of accidents. It will be important later on in the chapter, when dealing with the role of Julius Caesar Scaliger in the theses (section 5.1).

The relations between accident/actual inherence and quantity/extension in place are treated by the regents in the same way. They hold that actual extension must always be predicated of quantity: when it comes to physical bodies, it is always possible to pair off actual extension with extension in place and conversely potential extension with extension within itself. King 1612 is more sympathetic to a Catholic Scholastic phraseology when he writes that we must distinguish between extension of parts among themselves and extension in place: yet, his conclusion is that extension without being extended in place is not real extension, with a reasoning similar to Baron 1627 on actual inherence and accident. The shared view seems to be that extension in place is part of the essence of quantity; in particular, quantity provides matter with extension in place, not simply with extension. To obtain a body actually extended in place, quantity is all that is required. There is no need for a further actualisation of the internal extension. To underline the similarity with the question about accidents, Mercer 1632 uses the expression 'aptitudinal extension' instead of 'potential extension'.[16]

The link between quantity and extension in place is so strong that every quantified body is per se extended in place. The relation between quantity and extension in place is one of identity: it does not occur that quantity is without the qualification 'extended in place' because 'extended in place' is part of the definition (therefore of the essence) of quantity.

Mercer 1632 is the only regent to name the theory of Transubstantiation by reference to '*Metusiastae*', the 'proponents of μετουσία', the Greek name for Transubstantiation. The problem addressed in these passages by the regents is the impossibility for an extended body to exist without its actual extension, as it is required by the presence of Christ in the host. This remark inevitably leads to the question of the relation between extension and place.

As a conclusion of sections 2 and 3 of this chapter, we can argue that: 1) accidents cannot exist without a substance, because their essence is to be in a substance (*inesse*). According to the regents, Catholics want us to believe that in the miracle of Transubstantiation accidents are preserved without their substance; 2) quantity is an accident, thus it cannot exist without its substance. Furthermore, quantity is interpreted by Catholics as a 'quasi-substance', in which other accidents inhere. Even if this is the case, accidents inhering in quantity cannot be without quantity. But regents argue that quantity without its actual extension in place breaks this principle; 3) quantity must always be actually extended, because extension in place is part of its essence. And this is rejected by Catholics, who

[16] See also, for example: Forbes 1624, TP XVII; Stevenson 1625, TL XI; Armour 1635, TL VII; Wemys 1635, TP VII.

claim that Christ's body can fit in the host because his body does not have extension in place, though it retains its internal extension and proportions.

4. Quantity and place

The final concern regents have about the traditional account of Transubstantiation regards the theory of place implied by the real presence of Christ in the host. In order to make sense of the words of the Gospel, *hoc est corpus meum*, Catholics interpreted the corporeal presence of Christ as referring to the whole substance, which thus retains all its qualifications but one, inevitably, the extension of its parts in place. *Extensio partium in ordine ad locum* is thus considered as not identical with the extension of Christ's body, therefore it is separable from it without any changes occurring in what it is. The philosophy behind Transubstantiation is made coherent with the nature of the miracle.

Regents deploy against the core doctrines of this philosophy precise arguments aimed at showing its philosophical inconsistency:[17] the theory of place is simply derived from quantity as intrinsically extended in place.

4.1 Quantity and place as independent

To understand better the positions of the regents, a few remarks about the Catholic version of the problem are in order. I shall again follow Eustachius and Suárez, both of them for their clarity, and the latter on account of the direct references to him made in the theses.

The first concern is about *impenetrabilitas*, the power of quantity to resist the presence of another substance in the same place. The presence of the enduring accidents of the matter of bread and wine raises the question about their impenetrability with respect to the incoming substance of the body and blood of Christ. Eustachius makes a clear distinction between active and passive impenetrability. He affirms that:

[17] I am clearly not concerned with the theological rejection of Transubstantiation, which follows patterns different from what we read in the *Theses philosophicae* and with which it is arguable that all regents agreed.

> duplex esse munus quantitatis respectu loci; nempe locum replere, et ab eodem loco quodvis aliud corpus removere: Quod posterius praestat quantitatis penetrationi obsistendo, non quidem active, sed negative [...] Hoc autem posteriori officio privatur quantitas, cum duo corpora in eodem situ et loco ponuntur. [*SPhQ*, III, I, III, III]

In a now familiar way, Eustachius picks out what really belongs to the essence of quantity and what does not, to establish what must remain during the conversion. The distinction utilises the notions of activity and passivity, *'locum replere'* (filling a place) and *'aliud corpus removere'* (removing another body [from the same place]). The latter is said to belong to quantity only *negative*: it is not something that quantity does per se, but only something that follows from what quantity does per se, which is filling a place. Eustachius then concludes that it is this negative power which is subtracted from quantity during the conversion, when two bodies are placed in the same place.

Suárez openly states the connection between place and Transubstantiation:

> Quamvis autem Deus penetret duo corpora in eodem loco, non reddit illa non quanta, nec ex duobus quantis facit unum quantum, sed servata distinctione quantitatum constituit ea in eodem spatio. Sic ergo, licet Deus corpus bipedale constitueret in spatio pedali, non per condensationem, sed per partium penetrationem, non redderet illud minus quantum, neque duas partes in unam redigeret, sed in eodem spatio eas collocaret, quod longe diversum est. [...] Nego tamen substantiam sic constitutam in spatio indivisibili non fore quantam, nam corpus Christi quantum est etiam in sacramento, licet sit etiam in punto indivisibili. Et ratio est quia, ut dixi, quantitas non est actualis extensio in spatio, sed aptitudinalis, et hanc retinere potest corpus, etiamsi actu non sit in spatio extenso. [*DM*, 40, II, 22]

It is clear that both Eustachius and Suárez are here dealing with a non-natural occurrence from the expression *'Deus penetret duo corpora in eodem loco'*. In the normal course of nature impenetrability is a constant attribute of quantity, only a direct act by God can make two different things occupy the same place. Thus, the solution is that the quantities of two things occupy the same place; alternatively put, the quantity of two things, one of which is penetrated by the other one, is not eliminated, nor does one single quantified thing emerge from two distinct quantified things. What happens is that God can constitute these two things in the same place, while preserving the distinction of their respective quantities. The agreement with the dogma is complete because quantity can preserve the body, the whole body. The particular existence of this quantified body without its natural extension in place is that of an indivisible point, what has been called by some scholars 'ghostly matter' because it is all there yet it is deprived of its extension in place.

Two corollaries of this position are: 1) extension is not essentially measurable; and 2) the simultaneous presence of two bodies in the same place is not contradictory.

4.2 Regents' rejection of 'ghostly matter'

On this matter more than on others, regents refer directly to their chief opponent, Suárez. His theory of matter as shrinkable to a single point while retaining all of its qualifications,[18] thus its essence, was perceived as a very well-argued one, and unanimously criticised by regents. The fact that they mention Suárez directly shows that this theory was regarded as the best argued and clearest account offered by Catholic Scholastics.

Forbes 1624, one of the Aberdeen doctors, sharply states his two main concerns in one brief sentence: *"Docentes* [pontificij] *accidentia esse posse quamvis subjecto non insint, et corpus extensum, loco non mensurari. Quorum alterum accidentium naturae, alterum corporis quanti conditioni ita adversatur"* [TL XVII]. His objection is that the three fundamental philosophical premises of Transubstantiation (accidents not inhering in a subject, extended bodies not extended in place and, implicitly, quantity without its subject) are against both the nature of accident and extended body, thus are contradictory theories, not grounded in any essences of really existing things.

His colleague at Aberdeen, Seton, comments in 1637 on Suárez's theory of *simul praesentia* with the same words: *"Alii* [Suárez] *quantitativam individui corporis in pluribus locis simul praesentiam, corporis naturam plane evertere contendunt"* [TP IV].

Suárez's theory is seen as contradictory with respect to the natures of things involved. By extension, we can say that regents would consider in the same way a notion of place which is not the only extended portion of space where a body can extend itself.

Wemys 1635 has Suárez in mind when he writes that:

> Nullum corpus potest esse in loco definitive, nisi in eodem sit etiam localiter et circumscriptive. / Si corpus Domini non sit praesens in altari localiter et

[18] Apart from extension in place, as we have seen.

> circumscriptive, neque in eodem erit illocaliter, et (ut aiunt) sine modo quantitativo. [TP VII.1]

This is a direct reference to *DM* 51, VI, 2, where Suárez engages with the notion of being in place *definitive* and *circumscriptive*. Something is in place *definitive* when *"ita est alicubi, ut intra definitum spatium contineatur, nec simul possit extra illud naturaliter esse"*; while it is *circumscriptive "quando ita ibi est, ut sit tota in toto, pars in parte spatii quod occupat"* (*ibidem*). Suárez then affirms that:

> est autem subintelligenda negatio extensionis, nam alias etiam id quod est circumscriptive in loco, erit etiam definitive, quia non potest naturaliter simul esse in alio *ubi*; quo sensu illud esse definitive potest generice sumi prout distinguitur ab esse ubique, quod est proprium Dei. Ut ergo illa sit ratio specifica, subintelligenda est negatio, videlicet, ut res illa dicatur esse definitive in loco, quae licet non habeat in loco extensionem partium, intra certos tamen limites ita continetur, ut extra illud *ubi* naturaliter esse non possit. [*ibidem*]

Regents reject the possibility of *'negatio extensionis'* when it comes to bodies: a body cannot be in a place *definitive* (that is, a body cannot have its *ubi*) unless it is also in place *circumscriptive*: which means, when a body is somewhere in space, it cannot obviously be in two places at the same time, it must also be in the same 'somewhere in space' in respect of all its proportions and parts, the whole as a whole, the parts as parts. The qualification *circumscriptive* cannot be subtracted from *definitive*. According to the regents, Suárez's qualification of how something is in place is in contradiction with the notion of extension, because he does not include the extension of parts in place.

4.3 Scotus's rejection of the negation of Transubstantiation as applicable to the *Theses philosophicae*

M. Henninger, in his work *Relations*, investigates the criticism made by Scotus of Thomas's theory of relations: according to Scotus, it leads to the absurd conclusion of the negation of Transubstantiation. The argument is presented by Henninger as follows:

> i) before Transubstantiation, the accident of quantity inheres in its subject (bread), while afterwards this accident does not inhere in this subject;
> ii) but the same accident of quantity remains both before and afterwards;
> iii) suppose (i) [Thomistic view], i.e. that a real relation is really identical with its foundation;
> iv) then that the accident of quantity is identical with its relation of inherence in the bread [situation before conversion];
> v) THEREFORE, the quantity is really united to or informs the bread throughout Transubstantiation [consistent with ii and iv, inconsistent with i].[19]

Henninger reminds us that Ockham, despite opposing Scotus's theory of relations, found this argument so compelling that he admitted that at least in Transubstantiation accidents are not identical with their relation of inherence in their substances. It appears that Ockham finally rejected a position very similar to that of the regents.

It is noteworthy that the majority of the regents holds, as does Thomas, that a relation does not have a formal real entity, that a relation does not change the *relata* between which it occurs, and finally that a relation is identical with its subjects.[20] It seems that in the theses the few references to relations are often coherent with a Thomistic position, which according to Scotus fails to justify the dogma of Transubstantiation, as it appears from Henninger's account of Scotus. I argue that in general the regents are closer to Scotus than to Thomas, and indeed this is not a novelty in the seventeenth century, given the wide diffusion of the Scotistic school, which might even have outnumbered all the others.[21] If we may fail to identify a precise relationship between Scotus and the *Theses* in terms of being part of the 'Scotistic school', at least we are justified in saying that the *Theses*

[19] *Ivi*, pp. 75-76. At point (iii) the bracketed words are mine.

[20] For example: *"Potest Relatio ad subiectum accedere, vel ab eodem recedere, sine omni mutatione subiecti. / Ideoque cum nobilibus Philosophis Relationem a fundamento non realiter differre statuentes, nisi adversa ratione, in quorundam verba iuremus, formaliter acceptam realem entitatem nullam habere asseremus."* Adamson 1600, TL VIII.1-2; *"Relatio quum minimae sit entitatis, ut nihil reale ponit in subjecto, ita formaliter acceptam entitatem realem non habet. [...] Adeo ut haec sit distinctio rationis inter relationem et fundamentum, quod in conceptu relationis includitur terminus, qui non includitur in conceptu fundamenti."* King 1612, TL 14.1 and 5; *"Relatio non distinguitur realiter aut modaliter; sed sola ratione ratiocinata a suo fundamento proximo. et hinc est quod relationem subjecto advenire sine ejus mutatione, doceat Arist. 5. Phys. contex. 10."*, Barclay 1631, TL VII.4. Sibbald 1625 [TL IX-XII] and Seton favour the Scotistic theory: for example, Seton 1634, TL XL: *"Celebris, Thomae cum Scoto controversia est, eadem ne numero relatio, ad diversos numero terminos terminetur. Divi (hominis hic infirmi) Thomae affirmantis castra, succumbentis quippe, (Athletae alias insignis) deferentes, Scoti victoris vexillum sequimur."*

[21] *'The school of Scotus is more numerous than all the other schools taken together'*, Johannes Caramuel y Lobkowitz, quoted in A. Broadie, *A History of Scottish Philosophy*, Edinburgh, Edinburgh University Press, 2010, p. 1.

philosophicae reflect the eclectism of Scholastic philosophy of the early modern era, which is heavily influenced by Scotistic themes.

Thus, a Thomistic doctrine in a generally non-Thomistic context requires an explanation. I conjecture that the regents were motivated by their rejection of Transubstantiation. I am not arguing for an intentional allegiance to Thomism on this matter because of the rejection of Transubstantiation; but more simply, that because of the rejection of Transubstantiation, the regents found Thomas's theory of relations more appealing than others', and appropriated it without regard to other characteristic features of Thomism.[22]

On a theoretical level, regents seem to see the point of Scotus's attack on Thomas on relations, because 1) they do reject Transubstantiation; and 2) they share Thomas's theory of relations, which Scotus argues to be incompatible with Transubstantiation.

5. Protestant Scholasticism and Catholic Scholasticism

In the passages regarding Transubstantiation, regents usually do not mention any Scholastic source on their side, with the exception of Julius Caesar Scaliger and his *Exoticarum Exercitationum Liber XV de Subtilitate, ad Hieronimum Cardanum*, first published in 1551. All other philosophers are quoted with a view to criticising their position.[23]

The continuous presence of Scaliger throughout the theses, although not regarded as a fundamental source of inspiration, nonetheless sheds light on the historical question of the sources of Scottish Scholasticism. This question must be answered mainly on the basis of textual evidence of the *Theses philosophicae*, but also by doing a survey of library catalogues of the period: the sum of this information can give us the spectrum of the readings and of the philosophical knowledge available to the regents. In the case of Transubstantiation, no Scholastic authority seems to enjoy the favour of the regents, for the obvious reason that no traditional Scholastic philosopher ever attacked the dogma of

[22] I wish to point out that the regents distance themselves from Scotus on another important matter: the reference to inherence in the definition of accidents. We read in *DM*, 37, II, 5 that *"secunda sententia, praecedenti extreme opposita, est inhaerentiam nullo mode esse de essentia accidentis, neque actu neque aptitudine. Ita tenet Scotus, In IV, dist. 2. q. 1."* The regents include actual inherence in the definition of accidents: as Suárez reminds us, this is Aristotelian [*DM*, 37, II, 2].

[23] With one remarkable exception: Durandus de Saint-Pourçain. Murray 1628, TP I.5: *"Rectius Durandus, qui dicit materiam panis eandem manere in corpore Christi."* Regents look favourably at his claim that in Transubstantiation the matter of bread and wine is preserved through conversion. Suárez also quotes Durandus on this point, with the opposite intention. Despite this favour, regents cannot agree with the defence of Transubstantiation they find in Durandus's works.

Transubstantiation. On the side of the still much underexplored Protestant Scholastic philosophy, regents seem to focus mainly on Scaliger, even if not exclusively so.[24] From this individual case, it may be possible to draw some general conclusions about Scottish Scholasticism: philosophers in Scotland were very well versed in Scholasticism, and made extensive use of Scholastic terminology and theories, but this philosophy was of little help to a Protestant scholar who wanted to find a philosophical analysis of their belief in the rejection of Transubstantiation. It is plausible to suppose that regents had to develop their criticism much on their own.

Theological criticisms were abundant in the early modern era, but one of the key aspects of Protestant theology is precisely the attempt to do without the vast Scholastic philosophy added to it during the Middle Ages.[25] This does not mean that this enterprise was straightforward and accomplished from the very beginning; yet, at least in Scotland, a form of Scholastic theology is virtually absent in the seventeenth century.

Thus, the *Theses philosophicae* are an interesting example of Scholastic philosophy, intimately influenced by the faith of the Reformed Church of Scotland, but not on that account less Scholastic than equivalent Catholic Scholastic texts. More precisely, the theses develop the criticism of Transubstantiation as they do precisely because they are still a product of Scholastic philosophy.

5.1 Scaliger's *Exercitationes*: a possible source for the philosophy of the regents

Scaliger and a few other philosophers[26] offered to the regents extensive works in philosophy on the Protestant side of the debate, and this inevitably attracted their attention and favour. Scaliger above all others is always quoted with approval. Narrowing down his

[24] The second most quoted Protestant Scholastic is Bartholomeus Keckermann, Dutch philosopher renowned in the period for his textbooks on logic and natural philosophy.

[25] In Scotland for example, the theological theses by Andrew Melville of 1599, despite the title *Scholastica Diatriba de rebus divinis*, do not qualify as 'Scholastic' in the Catholic sense of a deep relation between theology and philosophy, and do not offer philosophical theories similar to those of the graduation theses. On Melville's theses, S. Reid, *Humanism and Calvinism*, pp. 191-193. See also Muller, *Post-Reformation Reformed Dogmatics*, *passim*, on the relation between Protestant theology and Scholasticism. Interestingly, R. Baron, in his *Philosophia theologiae ancillans* (St Andrews 1621), does not mention the philosophical criticism of the Catholic account of Transusbtantiation. Even if both Melville and Baron are influenced in form and contents by Scholastic theology, it seems that they clearly perceive the difference in themes and arguments between theology and philosophy. Particularly remarkable is a regent such as Baron, who does not fail to stress his agreement with Thomas, Scotus, Suárez, Ruvius and the whole Society of Jesus (1617, TP XXV).

[26] I am referring to Keckermann and, secondly, to Ramus. On Keckermann and Ramus in Scotland, S. Reid, *Humanism and Calvinism*, *passim*. In particular, pp. 259-264 for Keckermann.

contribution to the debate over Transubstantiation, two references are important. The first one is in Murray 1628:

> Ideo quia locus est spatium in quo necessario extenditur quicquid habet partes extra partes, sequitur quantitatem extendi non posse nisi extendatur in ordine ad locum: ut merito subtilis Scaliger subtilem Doctorem damnaverit, inquiens modum quantitativum non esse accidens per accidens, sed proprium proprio modo dictum. [TP VI.5]

The regent approves of Scaliger's remark, directed against Scotus, that the quantitative mode of a thing is not predicated as an accident (which can then be subtracted from its subject), but as a *proprium*, which, by definition, cannot be subtracted from its substance. We find this passage in *Exercitatio* V, 7, where Scaliger writes about the *modus quantitativus*, indeed not without irony, that *"Barbari nostri vocarunt id, quod rationem quantitatis dicere possumus. Non tamen ratio, qua quantitas est quantitas: sed est praescriptio corporeitatis in praedicamento quantitatis." 'Proprium'* is defined by Porphyry as *"what belongs always and exclusively to the totality of one species [...] always present in it by nature."* [*Isagoges*, 12, 17-19].[27] The proprium is something following from the essence of something, directly depending on it for its existence, yet not part of it. A traditional example is *risibilitas* predicated of man: it does not signify man's essence, but it follows from it and man cannot be without it. Scaliger's response to Scotus is precisely that quantity must be predicated as a proprium of its substance, not as an accident which is able to inhere in it and also able not to. In the competition for *subtilitas*, on this matter regents favour Scaliger over the Subtle Doctor.

The second reference is an implicit one, in King 1616, who paraphrases a similar passage in Scaliger's *Exercitationes*, V.6:

> Licet habitus, actio transiens et locus sint extra subiectum denominationis, extra tamen subiectum suae existentiae subsistere nequeunt. [King 1616, TL IX.4]
>
> Tametsi quod non includitur in definitione, abesse potest a definito, in definitione: non omne tamen abesse potest ab ipsa re definita. [Scaliger]

[27] '[E]'φ' οὗ συνδεδράμηκεν τὸ μόνῳ καὶ παντὶ καὶ ἀεί.' My translation.

The *Exercitatio* V quoted here deals with matter and void and related topics, such as extension and place. Scaliger is referring to place when writing that *"neque locum esse corporis necessarium, quatenus corpus est"* (*ibidem*). Indeed, the definition of body does not include that of place, indeed each one of the two can be defined without the other one. Nonetheless, Scaliger claims that if we go beyond the level of definition, it is impossible to find a body existing without a place, and (less evidently though) a place existing without a body. Concentrating on the latter (which is the object of King's passage), Scaliger and King agree on the fact that a place cannot be without the subject of its existence.

I take these passages to be coherent with the philosophy of the regents. In this debate, the most evident philosophical tool employed by proponents of the reality of Transubstantiation is the principle of separability, based on the real distinction between *res* (to be taken in the Scotistic sense, not necessarily as independently existing creatures). On the other side, regents exploited more (without naming it though) the possibilities of the *distinctio modalis*, the kind of distinction that exists between a thing and its mode, two things that cannot exist separately even if they have distinct essences. Thus, matter is not quantity, quantity is not extension, extension is not place, but none of them can exist without the other, even if each one of them can be conceived without the other. If we recall one of the conclusions of chapter 3, namely that matter is per se quantified, that conclusion is now better qualified by the analysis of Transubstantiation: regents seem to imply that matter is *per se* extended *in place*. In no passage do they explicitly refer to matter as body (which is for instance Zabarella's position), which is a theoretical step further in the direction of the attribution of positive powers to matter, and, ultimately, a dualism between soul and body. I think that regents did not go as far as that. Be that as it may, they claim that 1) matter is a substance, *quanta* per se; 2) that quantity is not conceivable if not extended in place; 3) and finally that this extended matter is per se *occupativa loci*.

I believe that the debate on Transubstantiation has shown some features which might be regarded as the most important contributions of the graduation theses in natural philosophy: 1) the definition of accident, revised to include *inesse*; and 2) the consequent move in the direction of the identity between proprium and accident.

6. Conclusion

I shall end with a brief statement regarding the ground for the theory of substance in the *Theses philosophicae*, which anticipates the analysis of the reception of Aristotle, in section

2 of the Conclusions. The analysis of Transubstantiation enables the regents to expound their theory on central themes of natural philosophy, logic and metaphysics, with repercussions in all their philosophy. The principle which unifies their approach originates in the definition of accident combined with the relation that an accident has with its subject. The context seems to be an overlapping of logic and metaphysics: this is not illicit in Scholasticism, because of the original identity between a definition and the essence of something. The case of prime matter is instructive: prime matter is actual (not formally actual) since it has an essence. The presence of an essence is enough to make prime matter a 'something', thus it is enough to make it actual. In this context, and indeed the regents show this little terminological ambiguity, talking of 'substance' is equivalent to talking of 'subject'.

The general approach in the regents' rejection of Transubstantiation is set when they hold that an accident cannot have an existence separate from its substance. Their view almost inevitably leads to the negation of the separability of quantity from matter, extension from quantity, place from quantity (this last point being more controversial, since place is not an accident of extension). In order to claim the reality of Transubstantiation, Catholic Scholastics are drawn in the opposite direction, allowing for the separability of accidents from the subject.

A corollary of this theory is the identity of an accident with its relation of inherence in its subject: if an accident is the same as its inherence in its subject (say, if extension is the same as its inherence in quantity), then this accident cannot be separated from itself, and consequently cannot be separated from its subject - preserving the distinction between accident and subject, because it is not identical with its subject, only with its inherence in the subject.

How far did the regents go in following this train of thinking? If the standpoint from which we look at the *Theses philosophicae* is the so-called modern philosophy, then the resemblance of their theory of matter with the notion of *res extensa* (despite unbridgeable differences) seems convincing.[28] My aim is not to impose a comparison which was clearly

[28] R. Ariew, in his *Descartes and the Last Scholastics*, Ithaca - London, Cornell University Press, 1999, part I, chapter 2, remarks the importance of Scotism in the Scholasticism of the sixteenth and seventeenth century, and raises the question of the absence of Scotism in scholarly works on the period, notably Gilson's *Index*. Ariew lists seven points on which Thomas and Scotus disagree: 1) the proper object of the human intellect, 2) the concept of being, 3) the human compound, 4) prime matter, 5) the principle of individuation, 6) space and 7) time and motion (p. 46). Ariew claims that *"Descartes leans toward Scotism for every one of the Scotist theses, as long as they are at all relevant to his philosophy."* (p. 55): I believe that the same can be said about the graduation theses. The difference between Thomas and Scotus on prime matter is particularly important in relation to the graduation theses and, as I suggest here, it is probable that the acceptance of Scotism paved the way to the reception of Descartes in Scotland. Regarding prime matter, D. Des Chene (*Physiologia*, p. 86) believes that between Descartes and the Scholastics *"the difference is this: the Aristotelians believed that God, according to his absolute power,*

unknown to the regents: the majority of the *Theses* were written before Descartes published his theory. The analysis must then be conducted within the limits of Scholasticism in order to get a fair impression of how "new" the philosophy of the regents was. Regents performed a substantial reinterpretation of key Scholastic doctrines: for instance, it is not completely accurate to say that *"that unintelligibility* [of the notion of an accident existing apart from its substance] *is* [...] *a by-product of the struggles of the new sciences against the Schools"*[29] because the regents claimed this unintelligibility and developed an answer to this problem still within a Scholastic framework. This opinion makes sense only if we take Scholastic philosophy to mean Catholic Scholastic philosophy. The regents could not accept the philosophical consequences of the belief in the dogma of Transubstantiation. Thus, they formulated a Scholastic philosophy whose specific Reformed character is, I argue, well exemplified by the theory of the actual inherence of an accident in its natural substance. The *Theses philosophicae* are an example of Reformed Scotistic Scholasticism.

could allow matter to subsist without quantity, while Descartes did not." Now, the choice of only Catholic Scholastic sources leads Des Chene to this conclusion: we have seen how the regents do not believe that God can allow matter to exist without quantity; and, in general, that there is no matter without quantity, and no quantity without actual extension in place. I argue that a deeper understanding of Protestant Scholasticism can shed light on our understanding of the Scholastic influence on and relationship with Descartes.

[29] D. Des Chene, *Physiologia*, p. 132.

Part II, chapter 1

Motus: general features of movement

Natural substances are composed of form and matter: form is the principle of actuality, prime matter is the principle of potency. All natural changes occur in virtue of the openness of prime matter towards form, form which can never fully actualise the appetite of prime matter. Even when informed, prime matter always retains the possibility of being informed by a form different from the present one. The relation between a form and its respective portion of matter is never necessary, therefore that portion of matter can be related to other forms (that is, informed) and will probably come to be so. This is why prime matter is said to be the root of becoming and ceasing-to-be.

The realm of 'nature' is limited by Scholastics precisely in terms of the notion of *motus*: everything which is 'natural' is *in motu*, and conversely everything which is *in motu* is 'natural'. The term *motus* is commonly translated with 'movement': I do not wish to engage with this commonly accepted translation, but I think that the semantic field of the word 'movement' as we intend it today may lead us astray from the original Scholastic context. In Scholastic natural philosophy *motus* refers to any change taking place within a substance or to a substance: a man's hair changing colour is a change in the category of quality, taking place within a substance, while a man walking is a change in the category of place, occurring to the substance 'man'. In today's terminology we rarely, if ever, describe the former change as a 'movement' and we tend to treat 'movement' and 'local movement' (locomotion) as equivalent. We no longer share the Scholastic worldview in natural philosophy, and hence, I think, this shift in meaning of the term 'movement'.

To Scholastic eyes, our contemporary meaning of movement is then restricted to one aspect only, namely, the changes occurring in the category of place (*ubi*). Traditionally, change is also in the category of quality (a change in a property of a compound depending on form), in the category of quantity (a change in a property depending on matter) and in the category of substance (the change of the whole substance). All these changes are movements. It appears that movement relates to categories, but is not in a category: movement is the process in virtue of which substances change; ultimately, movement is the very substance while changing. It is not a perfect (physical) act, so it has no proper category.

The *Theses philosophicae* expound a Scholastic doctrine of natural movement. Regents seem to agree with what we may call a standard account of movement which is proper to Scholastic natural philosophy, and which I summarize in three points: 1) movement is the process of change undergone by a substance; 2) it is structured and explained by act and potency; 3) it is a directed process, in which it is always possible to identify a *terminus a quo* and a *terminus ad quem*.

Other features are corollaries of these three points: for instance, the talk of 'natural places' and the role played by the 'agent' in causing natural movement. In principle, Scholastic natural philosophy accepts natural places to explain the perceived directedness of the movements of the four elements (earth, water, air, fire): earth and water are perceived to go downwards and air and fire to go upwards because their 'natural places' are respectively down at the centre of earth/universe and up at the first sphere, that of the moon. The first couple of elements is then 'heavy', the second 'light'. This theory will be deployed in chapter 2, during the analysis of the movement of heavy and light bodies.

When it comes to the role of the agent in causing the movement, it is a Scholastic principle that *omne quod movetur ab alio movetur*, 'everything which is moved, is moved by something else'. In principle, there is no such thing as an essential state of movement proper to bodies: all movements are directed, which means essentially limited to and ended by the acquisition of their end. The end of movement is *quies* (rest). Thus, every movement requires a cause, an agent. These general features will play a major role in the analysis of celestial movement, in chapter 3. A strong tradition has it that the sublunar world (the world of material substances, of the four elements and of natural corruption) is essentially different from the celestial world; that the components of celestial compounds are not the same as those of sublunar compounds. I shall argue that regents do not go beyond this traditional distinction, and propose an interesting reinterpretation of the principle *omne quod movetur, ab alio movetur*.

This chapter will focus first on the definition of movement and give a preliminary account of the relation of movement to its terminus, and secondly on the question of the relation between movement and the Aristotelian categories.

1. Definition of movement

Movement is so important in Scholastic natural philosophy because the very realm of what is 'natural', and therefore the object of natural philosophy, has to be accounted for in terms of the notion of movement. All substances constantly change, which is equal to saying that all substances are in constant movement. Movement is a necessary consequence of the materiality of substances, in fact the world of materiality appears to be the world of substances in constant movement. Scholastics go one step further than this claim when they attach a normative interpretation to 'nature': only substances in movement are properly called 'natural', and natural philosophy only deals with substances included in this notion of 'nature'.

The notion of movement is present in the very definition of nature, as it is traditionally taken from the works of Aristotle. *Natura* is famously the inner principle of movement of bodies: 'inner' because it must be proper to the body and not external to it; 'principle' because it must be the physical origin and explanation of the movement. This notion of nature appears to be normative: it is true that all substances move in virtue of their nature as principle, but it is also true that they move according to their nature as normative for their movement. A substance cannot naturally do anything which does not follow from its nature: otherwise stated, every substance behaves according to its nature.[1]

While introducing this theory of movement, it must be remembered that nature intended as normative of movement does not imply that Scholastics grasped the modern idea of physical law. In a metaphysics of substance, physical regularities are understood in terms of natural genera, not in terms of natural laws. A natural substance behaves according to its nature, which is what a substance is: substances belonging to the same species will consequently behave in a similar way, while substances belonging to the same genus will behave in proportion to the degree of similarity between the same substances in the genus. The Aristotelian theory of the immutability of natural genera is the final warranty of the universality of the notion of nature.

This is the only sort of normativity that we find in the natural world considered per se. In fact, a law in Scholasticism is always a law thought of by a mind, or a law present in somebody's mind: the *Theses philosophicae*, for example, accept the notion of *lex naturae* in moral philosophy, taken to be the law given to creatures (in particular to rational ones)

---

[1] This is another instance of the principle *operari sequitur esse*, which is ultimately the warranty for the regularity and the intelligibility of the natural world. The principle claims that the behaviour of a substance indicates the being of the substance, what a substance is. Therefore, our knowledge of the behaviour of a substance leads us to the knowledge of the nature and essence of a substance.

by God. Being God-given, this law is not a positive law: it is a natural law, because it mirrors how things are, and respecting this law is equivalent to respecting the nature of things. Thus, natural law is not a mind-independent set of norms which affects natural bodies, enables predictions of their behaviour, or is in any way mathematisable. It is a matter of dispute whether Scholastic philosophy is potentially open to the modern idea of laws in nature. It is however commonly acknowledged that a standard version of Scholasticism does not show any sign of such a deep and revolutionary shift of perspective.[2]

The normativity of nature implies the directedness of natural movements. When dealing with the relationship between form and matter, form has been qualified also as the 'end' of matter: matter has an appetite which is satisfied by form. The process of information is a movement, whose end is form. As we will see, it is debated whether form is equivalent to nature, or whether matter must be included in nature as well.[3]

The importance of movement is also highlighted by the different ways in which the subject of natural philosophy is usually expressed: Scholastics may offer a variety of answers, such as *ens mobile*, *corpus mobile*, *corpus naturale*, *mobile qua mobile*, *ens in quantum mobile*. In Scholastic philosophy, the accurate definition of the subject of a branch of philosophy is not a secondary task. Aristotle declared that each discipline has its proper subject of enquiry and that each specific discipline must follow its own rules, which are in a way dictated by the object itself. Thus, the subject of natural philosophy is, in general, the natural substance as it undergoes movement/change.

1.1 Movement as way, tendency and flux

John Case in his *Epitome in Octo Libros Physicorum* (Oxford 1599) chooses the formula *ens mobile*. He follows Thomas Aquinas's words in the commentary on the *Physics*:

> Et quia omne quod habet materiam mobile est, consequens est quod ens mobile sit subiectum

[2] See W. Ott, *Causation and Laws of Nature in Early Modern Philosophy*, Oxford, Oxford University Press, 2009. The author claims that *"the notion of law in this contemporary sense is alien to the Aristotelian family of positions. Where the notion does appear, it is in the context of a divine command theory of ethics"* (p. 21). Ott also investigates the position of Suárez, who claims that God concurs in secondary causation 'by an infallible law' (*DM*, 22, IV). Yet, Suárez only claims that God acts in a lawlike way; therefore, Descartes appeal to laws as secondary causes breaks with Scholasticism and it *"is a decisive point in the history of mechanicism"* (pp. 52-53).

[3] See below, chapter 2. If nature is what a substance is and does, the question whether form alone or form and matter together determine the nature of a substance is the same as the question of what a substance is.

> naturalis philosophiae. Naturalis enim philosophia de naturalibus est; naturalia autem sunt quorum principium est natura; natura autem est principium motus et quietis in eo in quo est; de his igitur quae habent in se principium motus, est scientia naturalis. [*In octo Physic.*, I, l. 1, n. 3]

Thomas can be considered as representative of standard Scholasticism regarding this theory, despite all the terminological differences with other Scholastics. The starting point is that everything which is material is mobile; therefore, the mobile being is the subject of natural philosophy (conclusion 1). Natural things (*naturalia*) are those things whose principle is nature, which is the principle of both movement and rest (*quies*); therefore, natural science deals with those beings which have the principle of movement in themselves (conclusion 2, qualification of conclusion 1). Even if the stress here is on movement, because movement is a natural state of mobile substances, it must be recalled that all natural things also have in themselves their principle of rest, which is of the exact same nature as the principle of movement. What makes something move, for the same reason eventually makes it rest.

The work by John Case is particularly interesting because it is geographically and chronologically close to the graduation theses. Case was an eclectic Aristotelian who lived and wrote in England in the sixteenth century.[4] The accessibility of his *Epitome* makes it a perfect work for the representation of a commonly accepted theory of movement in Scholasticism later than Thomas's. Case's definition is: *"actus entis mobilis in potentia quatenus fiat tale"* (chapter 10, *De motu in genere*), a slight rephrasing of the famous Scholastic definition *"actus entis in potentia quantenus in potentia est"*: movement is the act of a mobile being in potency insofar as it is in potency.[5] What Scholastics mean by it is that a being moves when it is in potency towards some end which it eventually reaches, and its 'being in movement' is precisely this 'being an act of a potency in quantum in potency'. Movement is the name of the passage from potency to act while still being in potency. When potency is actualised, movement is over and the end is reached. The difficulty inherent in defining something as 'act of a potency' is evident, but Scholastics

---

[4] John Case (1540-1600), philosopher. BA and then fellow of St John's College, Oxford, in 1568. MA in 1572 and MD in 1590. He published a number of philosophical works, mainly commentaries on Aristotle, including: the *Speculum moralium quaestionum in universam ethicen Aristotelis* (1585), the first major publication by the Oxford University Press. His natural philosophical works, *Lapis philosophicus* and *Ancilla philosophiae, seu Epitome in Octo Libros Physicorum* (1600), are his latest works. Case was an influencial lecturer in Oxford with sympathies for Catholicism, and contributed to the Aristotelian revival of the late Renaissance time. *DNB*. The main text on John Case is C. B. Schmitt, *John Case and the Aristotelianism in Renaissance England*, Kingston, McGill-Queen's University Press, 1983.

[5] For a survey of the different versions of this definition, D. Des Chene, *Physiologia*, p. 26, footnote 11.

are the first ones to consider the notion of movement obscure and difficult to define. Case adds some qualifications which make the definition clearer: 1) movement is an act, *inchoatus* (begun with, sketched, outlined), not a perfect act, since only form is a perfect physical act: in movement, a complete form is not yet attained; 2) the act is of a real being; 3) the act is of a mobile being, a being apt to move; and finally 4) the act is of a mobile being which is in potency towards something (ch. 10).

Among other definitions, Case favours this one as the clearest and most inclusive. Regents do agree with this traditional definition, even if their debate is mainly over a different definition of movement:

> [motus] est acquisitio ipsa, et tendentia ad formam, cujus natura adeo mobilis est. [Forbes 1623, TP VIII]

> [motus] est tendentia mobilis ad formam, et via inter duos terminos. [Baron 1627, TP IV.5]

The notions of *via*, *tendentia*, and *acquisitio* are recurrent in the graduation theses.[6] Regents seem to agree on the general idea that a movement is a process from the terminus-from-which to the terminus-to-which. It seems that the three terms *via*, *tendentia*, *acquisitio* are regarded as synonymous.

The broad debate is about the opposed notions of *fluxus formae* and *forma fluens*: is movement the flux of form, or is it the form itself while 'flowing'? Reid 1622 lists *fluxus* (flux) along with *via* and *tendentia*:

> [motus] nihil aliud est, quam via, fluxus, seu (ut loquuntur) tendentia de termino in termino. [TP X.1]

while Forbes 1623 and Barclay 1631 hold that:

> [motus] non est forma per se, nec forma pariter cum fluxu, seu acquisitione, sed est acquisitio ipsa, et tendentia ad formam. [Forbes 1623, TP IX]

> Motus non nisi imperite statuitur forma fluens: et inadaequate fluxus, seu successio formae. [Barclay 1631, TP 3.11]

---

[6] See also, for example: King 1620, TP VIII; King 1624, TP VII; Rankine 1627, TP VIII; Armour 1635, TP IV.4; Leech 1636, TP V.V.

The concept of *forma fluens* is of Scotistic origin.[7] Regents are almost unanimous in rejecting Scotus on this matter[8] and they side with the majority of late Scholastics in considering the movement as the flux of form. Barclay is an exception, for he says that the term 'flux' is inadequate to the explanation of movement. Scotus's theory is that movement is a succession of forms from the terminus taken to be the beginning of the movement until the final terminus. In the case, for instance, of a man getting from youth to old age, all intermediate steps are taken too: all intermediate forms are present in succession. Considered as a unitary movement (because the ageing man is the same man) we can say that the form of this man is flowing from youth to old age. We then have an enduring substance man whose 'parts' (different forms) are successive in time.

Forbes's passage seems to be a quite close quote from the Coimbrans' commentary on Aristotle's *Physics*, in III, c. 2, q. 1, a. 1: with the important evidence of the regent agreeing with the Coimbrans' solution of movement as acquisition and tendency. This passage is as follows:

> Motus secundum suam propriam rationem non est forma per se, nec forma pariter cum fluxu seu acquisitione, sed est acquisitio ipsa tendentiave ad formam.

Regents are disappointingly silent regarding their mutual differences in terminology. I believe that regents did consider little changes in the definition secondary, as they all imply a more fundamental agreement on movement as a 'process' from terminus to terminus. The late Scholastic debate over the *fluxus formae*[9] witnesses a substantial agreement between the regents and their continental colleagues. The general idea is that form cannot change, so movement is not a form, rather, it is the way towards form, the tendency towards form and the acquisition of a new form. Movement is not form, movement is something

---

[7] *Ordinatio*, II, 2, 1, 4.

[8] Stevenson 1629, TP XIII.1, claims that *"motus materialiter est forma fluens, formaliter fluxus formae."* This is a partial acceptance of Scotus, and perhaps an attempt to bring two different theories together. Unfortunately the passage is unique in the *Theses*, and Stevenson does not explain his claim any further. The idea that movement formally is flux of form is usually accepted; the claim that the matter of movement is a form 'flowing' towards the terminus is more debated. What Stevenson has in mind is, perhaps, a twofold account of movement: the form (that means, the reason) of movement is the flux of form, the matter (that means, the subject undergoing change) of movement is a flowing form.

[9] D. Des Chene, *Physiologia*, pp. 30-31. The author claims that the Scholastics prefer the definition of movement as 'flux of form' rather than 'flowing form', for the reason that movement is not *"the form itself acquired in passing"*, but rather *"the "way" or "tending" of that form toward another"* (p. 30): the regents are no exception. Regarding the translation of *fluens* with 'flowing', the author explains that form does not flow in the sense in which, for example, a liquid flows. 'Flowing' is rather a mode of existence of the form on the way, or towards the end of its movement. The difficulty and the regents' criticism of Scotus consist in the fact that a form is by definition unchangeable.

occurring to form; in order to avoid the perceived contradiction of a form changing, regents choose way, tendency and acquisition.

The Coimbra commentary on the *Physics* seems to be influencing the regents regarding the definition of movement. It is not surprising to find references to the school of Coimbra in the Catholic world, especially in those studia or orders which closely followed the philosophy of Thomas Aquinas; yet, it should not be unexpected in a Reformed environment like the Aberdeen colleges, in a Reformed country like Scotland in the 1620s. I believe that the general picture is one of ongoing and careful study of continental philosophers by the regents. As already mentioned when dealing with Transubstantiation, library evidence shows great attention devoted to Catholic Scholastic sources, attention which is not simply the collector's attention paid to relevant works: it is the attention of readers who actually made use of those works.[10] In matters of purely philosophical concern, the evident confessional gap seems to be carrying no weight.

This is an important point. While it might be supposed that regents limited their investigation of Catholic Scholastic philosophy to highly polemical and controversial doctrines for the purposes of the inevitable struggle between the opposed parties which followed the outbreak of the Reformation, the case of the definition of movement, which concerns on the contrary a rather neutral and a-confessional physical doctrine, shows how deep was the Protestant engagement with Catholic philosophy in Scotland. It must be said that the Coimbrans seem to enjoy a good reputation in natural philosophy, since regents from King's and Marischal in Aberdeen in particular commented with favour on another theory present in the Coimbra commentary: the exclusion of generation from the number of kinds of movements.

Prior to the analysis of how many different movements there are, and how many categories are directly involved, there is a question regarding the relation between movement and the termini, to which I now turn.

[10] MS M 70 in Aberdeen University Library is the oldest list of books in the library of Marischal College: it contains the list of books bequeathed by Thomas Reid in 1624 to the college, books which formed the core of the seventeenth-century library. Reid donated to the college an excellent philosophical library. Library catalogues are important because they show the range of sources at disposal of the regents, who are often quoting the relevant Aristotelian passages in the theses, but are rarely quoting their secondary sources. Thus, even if we should not exclude that books were available also via private acquisitions, the university library catalogue provides solid evidence. In Reid's list the following entries for Coimbra are listed: 1) *In Dialecticam Arist.* 1607; 2) *In 8 libros Physicorum* 1609; 3) *In libros de gen. et corr.* 1606; 4) *In libros de anima* 1609; 5) *In libros de coelo* 1606. All the texts were published In Cologne, *apud Bernardum Gualtherium*.

1.2 The distinction between movement and its termini

We have seen that movement is a process from a starting point (*terminus a quo*) to a final point (*terminus ad quem*) at which the movement ceases and the acquisition of the new form is complete. Regents reject Scotus's *forma fluens* on the basis that movement is not identical with form;[11] movement is not formally a form undergoing change, it is rather the change undergone by form (flux). This way, regents seek to make sense of the key reference to 'act of being in potency' that we find in the traditional definition of movement.

What about the difference (if there is any) between movement and its termini? King 1620 in theses VIII and IX makes the explicit link between movement as flux of form and the termini:

> Omnis motus successivus est, cum forma non simul acquiratur sed per partes: Et etiam continuus, quia est via et tendentia ad formam quae continuitatem importat.
>
> Cum motus nil aliud sit quam via ad formam quae absque successione concipi nequit, successio erit essentialis motui. [...]
>
> Motus non habet diversam existentiam ab existentia termini. Ergo nec diversam realitatem, cum nil aliud sit, quam termini acquisitio.
>
> Nihilominus motus cum sit formaliter fluxus formae et sua natura quid successivum, formaliter et essentia a termino distinguetur. Haec distinctio formalis sufficit, ut multa enuncientur de uno quae non de altero, imo attributa contradicentia. Nihil prohibet quo minus ea quae sunt eadem re, formaliter vero differunt; a se invicem separari et separatim subsistere possint.

King holds the theory that movement is the way to form, a way essentially continuous in space and time. The existence of movement is not different from the existence of the terminus, since movement is the acquisition of the terminus. The regent justifies this conclusion on the basis of the principle *sicuti res se habet quoad productionem, ita quoad existentiam*, which figures as the heading of thesis IX. So, movement and terminus are not really different. Now, King is not explicit about which one of the two termini movement is

---

[11] As we have seen in section 1.1, the most open rejections are in Forbes 1623, TP IX and Barclay 1631, TP 3.11, as well as in Baron 1621, TP 16. In general, the regents do not mention *forma fluens*, and the rejection of this theory is apparent from the endorsement of the opposite view (that of flux of form), or from the choice of a different terminology: way, tendency and acquisition. Stevenson 1629, TP XIII.1, speaks favourably of *forma fluens* even if he accepts it only in a material sense.

not different from: clearly, the more important terminus is the end of the movement (*ad quem*), because it is this terminus which is going to be acquired by the movement, while the *terminus a quo* (which is itself identified only when movements begins) is identical with the substance which is in movement.

We can then argue that silence on this matter implies that the regent is talking about the final terminus of movement. Yet, real identity and real difference are not the only two possibilities: it is true that movement is the acquisition of the terminus (so it is not really different from it) but it is also true that movement is formally the flux of form: the formal reason of movement is not the same as that of the terminus, therefore some degree of distinction between the two is required. According to King, the Scotistic formal distinction suffices in this case: the explanation rests on the idea that nothing prevents two things which are really the same thing (movement and terminus) existing one without the other, if they are formally distinct.[12]

This passage addresses some of the key aspects of the late Scholastic debate on the difference between movement and terminus. King's central idea is that a movement cannot be really distinct from its terminus, given that a real distinction only occurs between two things which can really exist one without the other. It is not the case that a movement, which is the acquisition of a terminus, can exist without its terminus. Thus, the answer must be found in a difference within real identity: real identity can happen (for instance, Coimbrans, *In Phys.* III, c. 2, q. 3, a. 2) also when the properties of two things are not entirely identical. It is in fact true: a difference in property does not entail that two things are entirely different.[13] This seems to be the case for movement and terminus. The majority of regents either deny that movement and terminus are really different or claim that they are formally different.[14] While the latter formula is to be preferred as more precise, the former is compatible with the latter.

I am not sure what sense to make of King's final remark that *"ea quae sunt eadem re, formaliter vero differunt; a se invicem separari et separatim subsistere possint"* if we take 'separatim' to mean 'real separation'. In this case, this remark would be in contradiction

[12] I find this remark a little troublesome: really separate existence is usually brought up by supporters of a real distinction between movement and terminus, such as Buridan. A possible interpretation of this remark is that, as Scotus would say, two formally distinct things enjoy separate existence in the intellect which think them separately; yet, King seems to exclude this when writing that two things which are the same *"separatim subsistere possint"*. In his theses of 1612, TP 6.III, King writes that: *"non est firmum illud Scotistarum: quorumcunque unum potest esse sine altero illa re distinguuntur."* The regents has a mode and its substance in mind: he seems to admit an independent existence which does not imply real distinction.

[13] Des Chene, *Physiologia*, p. 39. Quoting Coimbrans, Des Chene gives the example of a mode and the substance of this mode.

[14] For example: Reid 1622, TP X.2; Baron 1627, TP V.1; Stevenson 1629, TP XIII.2; Leech 1636, TP V.V.

with what has been said before, namely that the existence of a movement is not different from the existence of terminus. In fact, a formal distinction does not occur between really separable things. I also do not see what role this remark should be playing in the economy of the argument: the regent seems to prove the formal distinction by showing that the definition of movement as 'flux of form' is not the same as that of terminus.

Although dominant, the deployment of the formal distinction between a movement and its terminus is not the only route to a solution that we find in graduation theses.[15] This passage is taken from Wemys 1631:

> Motus est actus imperfectus ordinans et promovens subjectum ad actum perfectum, qui in ordine ad diversa aliam atque aliam induit rationem formalem.
>
> Realiter distinguitur a termino ad quem a quo sumit suam distinctionem specificam. [TP VI-VI.1]

The passage is regrettably too short to grasp Wemys's complete theory on the subject. It is perhaps significant that Wemys defines movement without the usual talk of way, tendency and flux. He stresses the act of movement in opposition to the act of the terminus: the act of movement being 'act of being in potency as it is in potency' is essentially different from the act of the terminus, which is physically perfect. The two formal reasons of the acts are different as much as 'imperfect act' is different from 'perfect act'. One more relevant absence in Wemys's words is the traditional philosophical argument by which real difference is usually established: God's powers. Two things are said to be really different even if they are not perceived as existing as two separate things so long as God could bring it about that they exist separately: in this case, Wemys could have told us that there is no contradiction in God sustaining the terminus without the movement, or vice versa. I believe that the reason for this absence lies in the afore-mentioned aversion that regents show for the appeal to God's powers in natural philosophy. This argument is considered an illicit appeal to something external to the realm of natural bodies, and therefore an appeal which ultimately endangers the autonomy both of natural philosophy as a discipline and also of the natural world as based on regularities to be found within the natural world itself. As I pointed out when dealing with Transubstantiation, Catholic philosophers in the seventeenth century on the contrary would accept this argument as philosophically relevant.

---

[15] For example: Sibbald 1623, TP 11; Lundie 1626, TP VII.

2. Movement and categories

Movement is not a substance: this means that movement is not a thing existing independent of a substance. It is then something happening to substances, and as with all such things the logical and metaphysical frame is that of the ten categories. Six of the categories are involved in the analysis of movement: four in a direct way, two only indirectly.

The four categories directly involved are: substance, quality, quantity and place. Any movement is traditionally thought to belong to one of these categories. As Wemys pointed out, movement is towards a terminus *"a quo sumit suam distinctionem specificam"* (TP VI.1): movement, being an imperfect act, is understood and categorised on the basis of the category of the terminus. This is ultimately why the *terminus ad quem* is prior to the *terminus a quo*, not in terms of existence, because no end of movement is possible without a beginning of movement, but in terms of reason: if we know the end of a movement, we know the category of movement and thus the kind of movement.

The two categories indirectly involved are *actio* and *passio* (action and passion). These two categories tell us whether a substance is being active or passive: for instance, 'walking' is an action, 'being touched' is a passion. The question concerns the relation of movement to action and passion: is movement properly posited in either of these categories? The regents address the debate mainly in reply to Suárez's claim that movement is identical with passion, except in reason.

Within the category of substance, two qualifications of movement are possible, one positive, generation (going towards a greater perfection), one negative, corruption (going towards a lesser perfection). *Generatio* is the formation of a new substance, *corruptio* is the dissolution of a previously existing substance. It appears that these two processes are the two sides of the same coin, according to the principle *generatio unius, est alterius corruptio*: from the corruption of a substance a new substance can be generated, and this is what happens in nature. Nonetheless, generation is prior to corruption by reason and existence, because in order to have corruption we must have something generated first.[16]

In the remaining three categories, 1) in quantity we have augmentation and diminution; 2) in quality, alteration; 3) in place, local motion, or locomotion (*latio*). Only quantity knows of a 'more' and 'less', while alteration and locomotion are presented as movement neutral to 'more' and 'less'. We will see how this is not entirely true of locomotion, as the

---

[16] Or rather, something created first. I address this point below, chapter 3, section 4.

qualifications *sursum* and *deorsum* (upwards and downwards) are not just accidents of locomotion but essential properties of movement (chapter 2).

2.1 Generation and movement

Regents debate over the inclusion of generation in the number of movements. Aristotle in book XI of *Met.*, 11, 1067 b 15-35 denies that generation is a movement. In fact, generation is the passage from a non-subject (a non-existing subject) to a subject: a non-subject cannot be in movement because it does not exist, thus the generation of a subject is prior to movement and is somehow the condition for movement to occur. This Aristotelian theory is the ground of the regents' discussions, as we are reminded by Lesley 1625, TP X,[17] who also, interestingly, quotes Scaliger's *Exercitatio* 290 as supporting the same theory. The authority of Aristotle who does not consider generation a movement does not convince the regents to endorse his view: indeed, it is a majority view but not at all the only one.[18] Here is how Forbes 1624 expresses the point:

> Forma omnis substantialis (cujus esse in indivisibili) per instans, ejus durationi intrinsecum, seu primum sui esse incipit, et ultimum sui esse desinit, quod est oriri, et corrumpi in instanti. [...] Unde errare eos patet, qui generationem substantialem, motui proprie dicto annumerant. Motus quidem est, in quantum motus a mutatione successiva et instantanea separat: at qua actum successivum ponit, ubi aliud post aliud, quod motui intrinsecum, vere motus non est, licet sumatur cum connexis alterationibus. Ita enim vel manet generatio, quae quia tempore non mensuratur, motus non est, vel ad duorum motuum confusionem in alterationem transibit: quo, quid absurdius? [TP VII]

The core of the difference between movement and generation is that the latter takes place in an instant, while movement takes place in time. It seems then that generation, which is only of a substantial form and so, of a whole substance, is rather called *mutatio* (mutation).

---

[17] William Lesley (d. 1654), university principal. Studied at King's College, and became regent there in 1617, sub-principal in 1623 and principal in 1632. We have one set of graduation theses by Lesley, the *Propositiones et problemata philosophica*, Aberdeen 1625. In 1638 Lesley signed the opposition to the National Covenant written by the Aberdeen Doctors, and was forced to resign from principal in 1640. *OG*, p. 54 and *DNB*.

[18] For the exclusion of generation from the number of movements, see also, for example: Carr 1617, TP VIII.2; Reid 1618, TP I-II; Forbes 1623, TP XIII; Ramsay 1629, TP III.12; Stevenson 1629, TP XIII.4; Wedderburn 1629, TP III.12.

Inevitably, generation brings about connected alterations (movements in the category of quality), but the regent warns us against taking these alterations to be the whole ongoing process. Forbes 1623, TP XIII, recalls the theory of Democritus as the traditional example of misunderstanding of alterations for generations, which Aristotle distinguished in a more careful way.

A similar view is held by Young 1617, in TP V:

> Motus est actus mobilis quatenus est mobile. Ergo generationi proprie dictae non competit definitio motus.

According to Young, the definition of movement is sufficient reason to discard the theory that generation is a movement: the conclusion is linked with the definition directly by 'ergo'. Generation cannot be a movement because movement is an act of a mobile as mobile: in generation instead, we have the coming-to-be of the mobile, not any sort of passage from potency to act of the mobile itself. It is remarkable that no regents mention the absence of contraries in generation as a fundamental difference between generation and movement. Aristotle himself first set out that a movement always occurs between contraries of the same species: according to the regents, a non-subject and a subject are contradictories, not contraries.[19]

The *Theses philosophicae* offer other examples of endorsement of this view, which is the most common one in late Scholasticism and which is also grounded in Aristotle's work. Yet, a small number of regents hold the opposite[20] view that what is going on in generation can be included in the definition of movement, as in Fairley 1619, TP V:

> Generatio sic actuat materiam ut non solum relinquat eam in potentiam ad formam, sed ut eam ordinet ad illam tanquam via ac tendentia ad eandem formam, et tanquam fieri ejusdem formae. [...] Ergo generatio stricte sumpta est actus entis in potentia quatenus est in potentia ad ulteriorem actum, qui est forma. [...] Definitio motus convenit etiam mutationibus instantaneis.

Now, generation is described by Fairley with the same words employed in the definition of movement: generation is an act of being in potency as it is in potency; it is a way and a

---

[19] *"Generatio essentialiter, est mutatio inter duos terminos, contradictorie distinctos."* Rankine 1627, TP III.6. For Aristotle's view, *Phys.*, I, 7.

[20] For example: Lundie 1626, TP VI; Barclay 1631, TP III.9; Mercer 1632, TP IX.2; Armour 1635, TP IV.6.

tendency towards form. If this were not enough, at the end of the thesis it is stated that mutations in an instant are movements. This passage is opposed to the standard theory of which Forbes and Young are representative. Even more interestingly it is also opposed to the set of theses written by the same Fairley in 1615 for the end of the curriculum of his previous class. In TP VII he writes: *"quod generatur non movetur. Generatio proprie dicta non est motus."* The textual evidence goes in the direction of a change of mind by Fairley, which happened some time during the four years after 1615, years spent in teaching undergraduates and studying. Such evidence is rare in the whole corpus of graduation theses probably because regents rarely took up teaching as a long-term job and they usually produced not more than two or three sets of theses each. It is then less likely to witness significant changes in the span of time of few years. The Arts Faculty of Edinburgh from around 1610 to 1625 is a good candidate for evidence of such changes, since philosophy teaching was conducted by the same regents, Young, Reid, King and Fairley, for quite a number of years. They produced some of the most complete sets of theses; of which, five sets of these are by Reid, five by King and three by Fairley, with no missing theses.

Fairley quotes Aristotle's *Physics* III, 1 at the opening of his 1615 TP VII, in Greek: H κινήσις ἐσιν ἐντελεχεία τοῦ δυναμεῖ οντος, ᾗ τοιουτον.[21] These exact words do not appear in the *Physics*.[22] Fairley probably intended to express in his own words Aristotle's thinking, which is not uncommon in the *Theses*. This definition of movement as 'act of being in potency qua potency' is the ground for the denial in 1615 that generation is a movement (as in Young 1617); while in 1619:

> Ad motum definitum libro tertio non est necessaria successio vel latitudo gradualis formae per eum acquirendae, ut mutatio dicatur convenire subjecto, quatenus est in potentia: sed satis est quod mutatio et forma sint duo actus, forma quidem perfectus, mutatio vero imperfectus et ad eam ordinatus, et eadem mutatio sit natura saltem prior forma. [TP, V.4]

Fairley is rewriting his own interpretation of the same passage of Aristotle. We are now told that the succession or gradual latitude of form (to be understood as the flux of form which is the movement) is not necessary for movement, also according to Aristotle's definition. Thus, what we call 'mutation' can be included in the definition too. Yet, Fairley

---

[21] I always transcribe the quotations respecting the regents' choice of accents, spirits and spelling of words. In the original, characters follow the style of sixteenth-seventeenth century printing.

[22] Fairley is slightly misquoting *Phys.* III, 1, 201 a 9-10: ἡ τοῦ δυνάμει ὄντος ἐντελέχεια, ᾗ τοιοῦτον, κίνησίς ἐστιν. W. D. Ross, *Aristotle's Physics*, Oxford, Clarendon, 1936.

feels compelled to identify a passage from an incomplete to a complete act, which is precisely what Aristotle denies in *Met.* XI, 11 in the case of generation. Fairley believes the solution to be that mutation is an incomplete act, ordered towards form which is a complete act, and that mutation is prior to form. In this way, granted the definition of movement, we also have the passage from incomplete to complete act, and not a passage from a non-subject to a subject.

The issue is not whether Fairley's solution in 1619 follows the letter of Aristotle or not. Two elements are evident though: Fairley deploys an Aristotelian theory, and Lundie 1626, who supports the idea that generation is a movement, exploits the same strategy as Fairley's.[23] The regent reads generation as falling under the definition of movement:

> mutatio enim materiae, a forma in formam actus quidam eius necessario est (per illam enim de potentia in actum educitur) non tamen perfectus (quippe non forma, sed ad formam via) ergo τοῦ δυνάμις ὄντος ἔντελεχεια ἡ τοιοῦτον, natura saltem actu perfecto prior. [TP VI]

In mutations too, a passage from act to act takes place; there is a 'non-perfect' act, which is the way towards form; therefore, there is an 'act of a potency as potency.'

[23] I treat the central topic of the reception of Aristotle in the graduation theses in the Conclusions, section 2. It appears that the regents substantially agree on the interpretation of Aristotle, in particular in relation to the adaptability of Aristotle to the Christian faith. This is not surprising, because this approach is central to Scholasticism as a whole, and the regents do not follow the path of some Renaissance Aristotelians, such as Zabarella or Cremonini, who, following the tradition of the medieval Faculty of Arts (in particular in Paris) read Aristotle as an alternative to Christianity. Nonetheless, differences among regents surface when it comes to the literal interpretation of some Aristotelian passages. Leaving the Christian interpretation of Aristotle aside for the moment, King 1616, TP II.1, writes that *"materia prima non est Aristotelis commentum"* (also quoted in Conclusions, section 2.2), while Stevenson 1625, TP VII, reading *Physics* II, 1, 193 b 9-12, claims that *"ut in artificialibus lignum se habet ad lectum, ita in naturalibus materia prima ad substantiam compositam."* It seems that Stevenson identifies in Aristotle a theory of prime matter, while Reid does not. Another example is taken from passages of the graduation theses in chapter 2, sections 3.1-2. Adamson 1600, TP VI and Lesley 1625, Problemata physica 9 read *Physics* VIII, 4, as claiming that, according to Aristotle, heavy and light bodies are moved by an external mover. Again, King 1616, TP XIII.4, corrects Adamson and Lesley by saying that Aristotle denies that inanimate bodies move themselves only in order to stress the difference between inanimate bodies and animate bodies. Another controversy arises on the matter of the interpretation of *Physics* VIII: I deal the regents' positions in chapter 3, section 4. The regents reflect two opposite approaches to Aristotle's passage:

2.2 Augmentation, alteration and movement

With regard to the relation between categories and movement, the debate over generation is the most relevant one but by no means the only one. One regent brings up the question whether augmentation can be properly called movement as well. This reference is unique in the *Theses*, yet it is remarkable for the philosophical arguments deployed in support of the claim that augmentation is not a movement. The passage is taken from a late Aberdeen set of graduation theses, Seton 1637:

> Accretio, motum localem non includit, cum illa momentanea, hic sit successivus. [TP XIV]

The regent finds support in another Catholic Scholastic, Ruvius, without referring to a specific passage. Seton's argument is the same as in the case of generation: movement always occurs in time, it is a successive and continuous process (successive by essence, continuous by accident) and this is the key qualification of movement, not a more general notion of passage from imperfect act to perfect act. If we compare this passage with Fairley 1619 for instance, these two regents hardly have the same theory of what is specific to movement, even if they agree on the terms of the analysis: act and potency, change in an instant, and termini.

The result is that local movement is not included in augmentation, because local movement and augmentation are different changes.

Now: the exclusion of local movement from augmentation does not itself mean that augmentation is not a movement. Local movement is not the only kind of movement, so Seton 1637 could be saying that augmentation and local movements are both movements, and simply different kinds of movement. I believe however that this is not what the regent had in mind. First of all, the opposition between 'change in an instant' and 'change in time' is usually deployed as a mark of the distinction between generation and movement. So, if we are to use the same opposition here, Seton is saying that augmentation is not a local movement because local movement is the only movement which really falls under the definition of movement. Augmentation is then another kind of change, similar to movement yet different from it.

Secondly, Seton seems to go against the traditional idea of local movement as the 'first' movement, which is prior to the other kinds of movement and, in some sense, their

---

according to some, like Sibbald 1623, TP 14-16, the contents of book VIII fall within the scope of metaphysics; according to others, like Wemys 1612, TP 13.I, they are part of natural philosophy.

foundation. It is true that Scholastics usually take local movement to be the archetype of movement,[24] but it is by no means the 'only' movement. And also, he seems to object to a less well-established but equally interesting idea that alteration is the 'first' movement. On these two points:

> Terminus (ad quem) sicuti speciem et distinctionem, ita nobilitatem motui absolute confert. Alteratio omnium motuum est praestantissimus, sicuti qualitas quantitate praestat. [King 1624, TP XIII]
>
> Principia lationis elementorum, posteriora sunt principijs generationis. Et consequenter ipsa Latio posterior est generatione in eodem, quamvis absolute in Universo, omnium mutationum prima sit. [Reid 1618, TP VIII 1-2]

Both positions exploit traditional arguments. King 1624 is basing his idea on the priority of quality over quantity, which implies the priority of form over matter; and on the qualification of movement given by the end of movement, qualification which includes some sort of 'nobility' of movement itself.

Reid 1618 on the contrary emphasises that local movement is not possible without generation, yet, generation is prior in respect of the temporal order but not by reason, because on a universal scale local movement is the first movement. Indeed, this is what Thomas writes in his commentary on Aristotle's *Physics*:

> Circa primum ponit duas rationes: circa quarum primam sic procedit. Primo enim proponit quod intendit: et dicit quod cum sint tres species motus, unus quidem qui est secundum quantitatem, qui vocatur augmentum et diminutio; alius autem qui est secundum passibilem qualitatem, et vocatur alteratio; tertius autem qui est secundum locum, et vocatur loci mutatio: necesse est quod iste sit primus inter omnes. Et hoc secundo probat sic: quia impossibile est quod augmentum sit primus motus. Augmentum enim esse non potest nisi alteratio praeexistat; quia illud quo aliquid augmentatur, est quodammodo dissimile et quodammodo simile. Quod enim sit dissimile, patet; quia illud quo aliquid augmentatur est alimentum, quod est in principio contrarium ei quod nutritur, propter diversitatem dispositionis. Sed quando iam additur ut augmentum faciat, necesse est quod sit

[24] *"Ille motus localis inter alios primus erit qui solus potest perpetuus esse et continuus. [...] propterea ille solus omnium erit primus et hoc motu movebit primus motor."* Toletus, *Commentaria in octo libros Aristotelis de Physica auscultatione*, Venice 1573, lib. 8, cap. 7.

> simile. De dissimilitudine autem non transitur ad similitudinem, nisi per alterationem. Necesse est ergo quod ante augmentum praecedat alteratio, per quam alimentum de una contraria dispositione mutetur in aliam. Tertio vero ostendit quod ante omnem alterationem praecedat motus localis: quia si aliquid alteratur, necesse est quod sit aliquid alterans, quod potentia calidum faciat esse actu calidum. Si autem hoc alterans semper esset eodem modo propinquum in eadem distantia ad alteratum, non magis faceret calidum nunc quam prius: manifestum est ergo quod movens in alteratione non similiter distat ab eo quod alteratur, sed aliquando est propinquius, aliquando remotius; quod non potest contingere sine loci mutatione. Si ergo necesse est motum semper esse, necesse est loci mutationem semper esse, cum sit prima motuum. Et si inter loci mutationes una est prior alia, necesse est, si praemissa sunt vera, quod prima sit sempiterna. [*In octo Physic.*, VIII, l. 14, n. 3]

In this passage Thomas is outlining a scale of movements, which justifies the pride of place given to local movement, and places alteration prior to augmentation. Thomas's arguments are strictly physical in this text, but I suppose he would not reject King's parallel solution of the problem.

2.3 Movement and the categories of action and passion

The principle *omne quod movetur ab alio movetur* is central to Scholastic natural philosophy. It is the premise for two of the most influential passages in the history of philosophy: book VIII of the *Physics* of Aristotle which proves the existence of a first motor and *Summa theologiae* I, q. 2, a. 3, where Thomas introduces the five ways to the affirmation of the existence of a first cause, which is usually called 'god'. Regents do endorse this principle, even if I think that they put forward an interpretation of it which strays from traditional Scholasticism. I will examine this interpretation later on, in chapter 3, section 4.

What is relevant now is that this principle directly entails the existence of an agent as cause of movement, and this relates to the Aristotelian categories. The agent acts on a patient (category of action) and the patient is acted upon by the agent (category of passion). Thus, despite the absence of the agent/patient distinction in the definition of movement, it is generally accepted that there is no movement without an agent and a patient. Movement is in the moved thing as in its proper subject (sometimes movement is called an *affectio* [affection] of the moved thing) and it is in the mover as in its principle. The movement has its beginning in its principle, the mover, and its realisation in its subject, the moved thing.

How does this activity of the agent (action) and the reception by the patient (passion) relate to movement? Regents are divided on the distinction between passion and movement:

> Actio et passio distinguuntur formaliter a motu per ordinem et habitudinem, haec quidem ad subjectum, illa ad principium. Motus in actione et passione includitur, vel tanquam quid superius et transcendens. [...] Aegre nobis persuadebit Suarez motum ut est actus mobilis non differre a passione, ne quidem per actum rationis; sed passionis et actionis nomina esse synonima. Creatio formalem rationem actionis transeuntis et motus, non autem passionis participat. [Barclay 1631, TP 3, 5-8]

> Non recedendum est a recepta Peripateticorum doctrina asserentium haes tria, motum, actionem, et passionem, inter se non distingui realiter, sed tantum distinctione rationis (ut vocent) ratiocinatae. Etsi actio et passio possint esse sine motu, ubicunque tamen est motus, ibi necessario adsunt et actio et passio. [Baron 1627, TP V. 1-2]

Baron and Barclay taught in the same years in St Salvator's College, St Andrews, so their disagreement is particularly revealing. As is often the case, Suárez is the target of the regents' attacks: the theory that passion is only different from movement *ratione ratiocinata* is peculiar to him and did not have great success in late Scholasticism. Suárez goes even as far as saying that movement belongs to the category of passion [*DM*, 49, II, 4].

A distinction of reason occurs between two things which are not formally and actually different which are nonetheless different in our conception of them. The qualification *ratiocinatae* entails that such distinctions in our concepts are not entirely ours, but have some ground in reality; if they have no ground, the distinction is *rationis ratiocinantis*. Baron does not refer to Suárez on the alleged passion-movement identity; yet, the regent holds Suárez's theory that movement and passion do not differ really or formally, but only by reason, and he even ascribes this theory to the whole of a vaguely defined 'peripatetic school'. The identity is not complete because action and passion are without movement, even if the contrary is not true.

Barclay opposes this view with an argument for the existence of movement without passion. According to the regent, movement, action and passion always go together in natural philosophy, but we cannot infer from this a difference less than the formal one because there is at least one case, creation, where there is movement without passion. Creation's formal reason is said to be of the same sort of a transeunt action (a transeunt action is an action whose effect is different from the action itself or from the cause of the action: in this case, creation is different from God and his creative act) but also of movement. This objection rests on the idea that movement in general is the passage from potency to act: otherwise, I do not see how it can prove what Barclay intends to prove. It is accepted that in creation there is no passion, because there is no passive subject (since the subject, a created being, is brought about in creation, it does not exist before the creative act) and nothing passive can be referred to God. If we follow Barclay's example, Baron should be committed to hold that creation also involves passion, because there is movement in creation and movement is always with passion.

Yet, if generation is not a movement, because it is a passage from a non-subject to a subject, creation is even "less" of a movement. In fact, creation is prior to generation in existence and by reason, since its antecedent is pure nothing, not simply a non-subject. These considerations seem to carry no weight in Barclay's example.

3. Conclusion

In this chapter I have investigated the general features of the theory of movement of the graduation theses. 'Movement' is a 'process' (a way, a tendency) from the *terminus a quo* to the *terminus ad quem*: that is, from one form to another form. This is why the regents call it 'acquisition of a new form', which takes the place of the present form.

The notion of movement is very general, since it includes all natural bodies. In Scholastic natural philosophy, a body is properly called 'natural' when endowed with a nature, which is the inner principle of the movements of bodies. Each body moves according to its nature, thus, different bodies move in different ways. We will see in the next two chapters two particular occurrences of this notion: the movement of heavy and light bodies and the movement of the heavens. In particular, celestial bodies are of a different nature from sublunar bodies, therefore some features of sublunar movement are absent, such as corruptibility.

Despite these differences, the features of movement investigated in this chapter have set the theoretical framework for an understanding of the theory of movement of the theses.

The main points are: 1) movement takes place in time. The regents exploit this feature to mark the difference between changes in time, properly called movements, and changes in an instant, properly called mutations (*mutationes*). Generation and creation are not movements because they occur in an instant. 2) Movement is predicated of substances, it is not a substance itself. The majority of the regents deny that the change in the category of substance is a movement, ergo movement can happen only in the remaining nine categories: in particular, in the categories of quality, quantity and place.

My enquiry into the general features of movement will be completed in some central aspects by the investigation of heavy and light bodies and celestial bodies: in particular regarding the finality of movement and the role of the agent.

Part II, chapter 2

Movement of *gravia* and *levia*

The movement of heavy and light bodies (*gravia* and *levia*) is a case of the fourth type of movement, local movement (*latio*, *motus localis*). Local movement is the acquisition of the terminus in the category of place (*ubi*). Since movements are specified by their respective termini, movements are also categorised by the category of their termini. Local movement is the only type of movement which has an external terminus, namely the place where the body is; contrary to the other types, whose termini are something of the body itself: quality, quantity and of course substance are internal to the body in movement. This characteristic will be important when highlighting the differences between local movement and other movements. Nonetheless, the *ubi* of a body is truly predicated of the body: place is also a relational notion, but first of all a categorial notion.

A heavy or light body is a body which is drawn by nature respectively towards the centre of the world, downwards, or towards the lower limit of the sublunar sphere, upwards. Inherent in this cosmology are the two doctrines of natural finality and natural place. The upwards and downwards movement is also explained by the causal power of the end (*finis*) of such bodies, naturally (thus necessarily) driven towards their end. The natural place is the end: a respective place in space where all bodies would cease any further downwards or upwards movements.

The structure of this chapter is then divided into: 1) the analysis of the notion of heaviness and lightness; 2) the analysis of natural places; 3) and the explanation of finality and movement of heavy and light bodies in terms of nature.

1. Heaviness and lightness

The Scholastic notions of heaviness (*gravitas*) and lightness (*levitas*) are foreign to our contemporary worldview. In our scientific language only the word 'gravity' has been retained, while taking on a meaning different from its original one. The other side of the Scholastic coin, 'levity' today only has a relational meaning: something is light only relative to something else which is heavier. Taken in a non-relational way, 'being light' does not mean anything. On the contrary, in Scholastic natural philosophy heaviness and lightness are positive properties of bodies: a body can be heavy or light absolutely speaking.[1] This means that these terms tell us something about how things are in themselves, not in relation to something else, or in relation to a scale of measurement. Relations and degrees are admitted, but only as relations and degrees among substances with different properties.

The background of this theory is the doctrine of the four elements:, which are, in order of heaviness: earth, water, air and fire. These elements are the fundamental components of every body within the sphere of the moon, and hence of every body which is subject to generation and corruption. Traditionally, Scholastic natural philosophy accepts the difference in nature between the so-called sublunar world and the heavens, incorruptible and eternal. Even when the distinction does not entail a difference in nature between sublunar matter and celestial matter, the sphere of the moon is always intended as the limit of the world composed of the four elements, with all the consequent properties.

In this chapter the relevance of the theory of the four elements is due to the grounding of heaviness and lightness of bodies in the proportions between elements in each body. Aristotle dealt with this cosmology in his *De generatione et corruptione*, usually referred to by regents in the Renaissance version *De ortu et interitu*.[2] The influence of this work in the *Theses* exceeds the scope of the analysis of movement: it must be noted that regents dedicate much attention to elements and their *mistio*/*mixtio* (mixture) in natural bodies, and also that much of the special physics (for instance, nutrition and theory of heat) is centred on the theory of elements.

Elements are the origin of heaviness and lightness. This means that a heavy body, say, a stone, is predominantly composed of heavy elements (in this case, earth); conversely, a

---

[1] 'Being cold' and 'being hot', 'being wet' and 'being dry' are similar cases: as in King 1624, TP XXI. A contemporary notion similar to 'absolute cold' could be the point of absolute zero: yet, the other side of the scale, heat, does not have a limit.

[2] In *Aristotle and the Renaissance*, pp. 86-87, C. B. Schmitt explains the origin of this alternative translation. *De generatione et corruptione* is the usual form in the Middle Ages; *De ortu et interitu* began to be preferred from the time of the Vatable translation of Aristotle, in 1519. Among others, Coimbra commentators choose this version. Cicero was the first one to introduce *ortus* and *interitus* into philosophical Latin, later to be changed by medievals into *generatio* and *corruptio*. In the sixteenth century, the Ciceronian translation is preferred, as more coherent with the idea of going back to a purer Latin than the one inherited from the Middle Ages. Regents too prefer this version, but I do not think that this alone can be taken as evidence of a 'Humanist' agenda in the Scottish universities in the seventeenth century: in fact, by this century this translation was somehow parallel to the medieval one. A similar point can be made about the use of Greek quotes from Aristotle in the *Theses*: these elements are a heritage of Humanism more than a sign of an enduring Humanist approach.

light body, fire, has a predominance of light elements. A sort of half-way case is a feather: it is a heavy body, because its natural movement is downwards, yet its behaviour testifies for a different elemental composition from a stone's. Regents usually see heavy and light bodies as opposed cases of the same movement, as is proved by the formulae '*gravia et levia*' or '*gravium et levium*'. In a way, explaining the behaviour of heavy bodies is also explaining the behaviour of light bodies. Despite this parallel, heavy bodies enjoy a privileged position in the theory, because in our experience the downwards movement is predominant.

A general picture of graduation theses shows that regents did not reject the Scholastic cosmology based on the distinction between sublunar world and heavens and between upwards and downwards as natural directions of different elements. In this chapter and in the next one on the movement of the heavens it will be clear that regents put this general framework to a test: a significant case is the set of theses of 1626 by Reid, who puts forwards a substantially revised version of the Scholastic theory of movement.

1.1 Definition of heaviness and lightness

A proper definition of heaviness and lightness is missing from the *Theses*. This is explained by the fact that these notions are taken to reflect a basic, non-theoretical fact from our experience, thus a starting point for explanation rather than a conclusion of an argument. Definition must provide an account of the essence of the defined thing: in terms of heavy and light bodies, regents do not see how this can be any different from saying that heavy bodies are heavy and light bodies are light. Speaking in terms of elemental composition does not convey a definition either, but simply a description. As late as 1629, Stevenson makes this point clear:

> esse gravius nihil aliud est, quam per naturam alteri substare, et esse levius est alteri superminere. [TP XXII.1]

The regent gives us a description of the behaviour of heavy and light bodies in relation to one another: heavy bodies are below, light bodies are above, by nature. As I said earlier, relation is included in the notion of heaviness and lightness, yet by *'per naturam'* Stevenson indicates that this behaviour tells us something about the nature of these bodies, about what they really are. So, by nature heavy and light bodies have their own place in the structure of the universe, and reference here is not made to natural place. This is due to the difference between, on one side, the actual structure of the universe and, on the other, how the universe would look if 1) the elements were the only existing thing, and 2) they were left free to attain their ends. This is a passage from Reid 1618 which follows the already quoted passage that local movement if the first of all changes:

> Elementa per gravitatem et levitatem primo Mundum suo ordine constituunt: deinde per primas qualitates in se invicem permutantur. Iure igitur Arist. ordinem servans naturae, prius in libris de Coelo, de gravitate et levitate disseruit: posterius in libris de Ortu, de quatuor primis qualitatibus. Elementa prius sunt mobilia ad locum, quam generabilia. Non solum simpliciter, et in universo, latio omnium mutationum prima est, sed etiam prior est generatione in eodem sicut in elementis apparet. [TP IX]

Reid holds that elements are essentially heavy or light and that they immediately structure the universe in an orderly way by finding their place according to nature. This is one of the rare passages in which *gravitas* and *levitas* are used as nouns: the usual phrasing favours *gravia* and *levia* because the adjective respects more the notion of heaviness and lightness as properties of substances.

One more aspect is important: downwards and upwards movements are types of local movement, and this is why local movement is said by Reid to be the first type of movement in general. This conclusion can then be attained in two ways: either by showing how local movement is implied by any other type of movement, or, as in this case, by means of a basic cosmology, in which elements by local movements immediately compose the universe in an ordered structure. This local movement is also prior to generation, because elements concur in originating the fundamental stuff (prime matter) which is itself prior to generation and corruption. An interesting view, which completes the account of prime matter in a way that seems similar to modern philosophy.[3] If it completes the account of prime matter, it surely does not substitute it, since Reid seems to be the only regent who holds this view. In the history of Aristotelianism, this passage hints at the long debate over which book between the *Physics* and the *De generatione et corruptione* is prior by order of knowledge and/or order of being.[4]

---

[3] Reid is putting forward a brief account of the organisation of the universe by heaviness and lightness which might remind us of Descartes' famous mental experiment in *Le Monde*, where the passage of the universe from chaos to order is explained by natural laws only. I believe that both Reid and Descartes consider these accounts as logical and not chronological, since the world was created by God instantly.

[4] On this aspect, E. Kessler, *Metaphysics or Empirical Science? The two Faces of Aristotelian Natural Philosophy in the Sixteenth Century*, in M. Pade, *Renaissance Readings of the Corpus Aristotelicum*, Copenhagen, Museum Tusculanum, 2001, pp. 79-101. One of Kessler's conclusions is that the different

2. Natural places

Elements get into union in a mixture, whose result is, with different proportions, the totality of natural bodies. These bodies are heavy or light in consequence of the proportions; and also behave according to their nature of being heavy or light. Heavy bodies move downwards towards the centre of the earth (which is also the centre of the universe, in a geocentric cosmology), light bodies go upwards, towards the sphere of the moon which is the first sphere of the heavens, and is the limit and first container of the sublunar world. The natural place of heavy bodies is the centre of earth, the natural place of light bodies is the upper limit of the sublunar world.

The notion of natural places is of Aristotelian origin, and up to the regents' time it seemed to account successfully for the apparent directedness of natural bodies. The movement of heavy and light bodies qua movement follows the patterns outlined for any type of movement: 1) it is from one terminus to another; 2) it eventually comes to a rest (*quies*); 3) it is the acquisition of a new terminus (a new *ubi*) by a mobile put into motion by a mover.

Point 2 concerns natural places; point 3 is the subject of the last part of this chapter, where I deal with the principle of movement of heavy and light bodies.

When a body in motion reaches its end, the movement is over: a new form is acquired, the particular potency triggered by the mover is now actualised and the body undergoes another movement. Rest is not an absolute achievement for sublunar bodies, it is always a relative notion: rest is relative to this or that particular movement. We appreciate again the importance of the idea of materiality as perennial principle of movement: in cases of generation, corruption and local movement, materiality is a potency never 'extensively' (*extensive*) satisfied by formal acts: that means, no form can turn material potency into a pure act. A body can be in complete rest only in its natural place, a state which is subject of speculation, not experience, since the actual structure of the universe does not allow for

approaches based respectively on the *Physics* and on the *De generatione et corruptione* eventually led to *"the modern distinction between natural science and philosophy of nature"* (p. 100), in the sense that the reading of the *De generatione et corruptione* provided the ground for a 'naturalistic' approach to natural bodies, as opposed to the 'philosophy of nature' of the *Physics*. I believe that the graduation theses do not fall in the categorisation deployed by Kessler for Renaissance philosophy. In fact, there seems to be no apparent shift between two different accounts of nature in the interpretations of the two texts. The natural body is explained in terms of substantial form, which determines the essence but which is also received in matter in virtue of a certain mixture of the elements: Kessler considers this approach as proper to medieval Scholastics (*ivi*, pp. 80-81).

such a polarization of elements, which would tear natural compounds apart. In Scholastic natural philosophy, the universe is constantly held together by the intrinsic rationality of its components and their arrangement: each body is made for a purpose, its particular nature (*natura particularis*) is to be understood within its general nature (*natura universalis*), which aims at the harmony and coherence of the whole. In the next chapter, the analysis of the movement of the heavens will inevitably draw from this cosmology: for the moment, this briefly sketched theory works as the background for the theory of natural places.

So, even if the particular nature of a heavy body dictates that it goes downwards, its universal nature is also affected by other principles at work: 1) the principle that everything which moves is moved by a mover; and 2) the famous Scholastic fear of the vacuum (*horror vacui*) which entails that all bodies always move in order to prevent the occurrence of a vacuum. These principles, the elements, the mover and fear of the vacuum determine the movement of natural bodies, which are usually called 'mixed' (*mixtum*). 'Mixed' because every movement is the result of: 1) the action of the mover, which triggers and gives direction to the movement; 2) the nature of heavy and light bodies, which drives them respectively downwards and upwards; and 3) the physical need for continuity and proximity of matter: all these together explain why bodies behave as they do. Rankine 1631 offers an insight into this complex doctrine:

> Sicut corpus grave, remotis impedimentis sponte descendit, ita ob metu vacui, aut turbatum ordinem universi ascendit, absque ullo extrinsecus impellente.
>
> Non magis naturaliter corpus grave ordinarie descendit, quam in hisce casibus extraordinariis ascenderet.
>
> Non dicitur corpus grave in his casibus contra naturam particularem, et secundum naturam universalem ascendere, quasi natura universalis esset quid distinctum et superadditum naturae particulari, sed potius secundum particularem, sed appetentem bonum universi. [TP XVI]

Provided that Rankine rejects the notion of universal nature as anything 'added to' the individual nature,[5] we can interpret universal nature as part of the individual nature in what pertains to the good of the universe (*bonum universi*): every body then reflects in itself the grander structure of the universe, and concurs to its preservation. Thus, an upwards

[5] I believe that Rankine's remark is another case of the theory that only individuals exist, coherent with the regents' theory of substance: thus, the so-called 'secondary substances' (the universals) and the universal nature of bodies are not something existing outside the individuals.

movement which appears to be in contradiction with the heaviness of a body is as natural as the downwards movement.

2.1 Natural places and *quies*

In Reid 1614, TP 3, we read that:

> Motus est perfectio non perfecti, sed perfectibilis: quies autem perfecti perfectio est, et cujus gratia moventur mortalia.

Generally taken, rest is more perfect than movement because actuality is more perfect than potentiality. Being the actualisation of a potency, rest is also the end of movement, and all 'mortal' beings (or, in other words, all natural beings) move towards rest. Rest, as actualisation of a potency, is the state in which natural bodies would be if they were not natural, that means, if they were not act and potency. It is then clear that rest is only provisional, relative to a particular movement. It holds true that if a relative rest is 'the perfection of something perfect', the rest following from the acquisition of a natural place is even more perfect than relative rest.

It is also accepted that:

> duo motus contrarii magis pugnant, quam motus et quies. Ergo motus motui magis opponitur, quam quieti. [...] Corpora subcoelestia moventur ut requiescant. In iis quies est finis, ideoque motu praestantior. Et cessare a motu praestantius, quam moveri. [Young 1613, TP 16]

Young states the connection between rest and end clearly: all sublunar movements are essentially directed towards rest, and this is why rest is more perfect than movement. In consequence of rest being the perfection of movement, two contrary movements are said to oppose one another more than movement and rest.

Natural places have the power of attracting and preserving (*vis attrahendi et conservandi*) their respective elements and bodies composed of those elements, and this is precisely what distinguishes them from place in general. Regents usually accept a traditional definition of place, taken from Aristotle's *Physics*. Among slightly changing definitions in other *Theses*,[6] Stevenson 1629 includes the powers of the natural place in the definition of place:

> Locus est corporis continentis terminus primus et immobilis, eiusque proprietates sunt attrahere ad se locatum, illud conservare, et continere. 4. *Phys.* 4. [TP XV]

This definition is almost word for word taken from *Phys.* IV, 4, 211 b - 212 a 20 ff., where Aristotle writes that place is the first immobile limit of the containing body and immediately after states the natural relation between contained bodies and place, divided into downwards and upwards, because the limited thing always goes together with the limiting thing. I believe that regents accept this Aristotelian account linking the definition of place with the doctrine of natural places: the two sides of the coin cannot be taken separately.

Natural places have powers that places in general do not have: elements (and the bodies they compose) tend towards their natural places by nature while they do not have any natural preference with respect to any one of the accidental places they move to. It might be said that a heavy body prefers to be somewhere in a straight line between where it is and the centre of the universe, rather than be anywhere above where it is. When a moving body reaches its new place, the terminus of this movement is a new 'whereness' (*ubi*). Regents have it that this 'whereness' is the intrinsic terminus of local movement, not the surface of the containing body:[7] this remark will appear in all its importance in the next chapter, when dealing with the negation of *resistentia medii* in the heavens. For the moment, whereness and natural place can be taken as synonyms.

Once a body has reached its natural place all its natural movements (downwards or upwards) are actualised, and reach a stop. The regents seldom talk about the state of a body in its natural state, because it is not a possible object of our experience: what we can say is a matter of deduction, not experience. Reid, in two sets of theses, 1614 and 1626 offers more insights than the other regents, who limit themselves to listing the attractive and preservative powers of natural places. In 1614, TP 3.7, he writes that:

> si manere in suo loco sit quiescere, omne corpus naturale sine exceptione quiescere potest.

[6] For example: Robertson 1610, TP 7; Bruce 1614, TP IX; King 1616, TP V; Baron 1617, TP XIX; Martin 1618, TP XXI-XXII; Reid 1626, TP VI; King 1628, TP VIII.

[7] Eustachius holds the same theory: *SPhQ*, pars III, tractatus III, disputatio II, quaestiones I-II.

The identity between 'remaining in place' and 'resting' is accepted, yet it is introduced by a conditional: *'si manere...sit quiescere'*. In 1626, Reid will revise this theory by making a distinction between the two terms of the identity.[8] One question is prompted by the concept of rest. Attaining a natural place is natural to bodies, rest in a natural place is the most perfect actualisation of the potency of movement: is this rest completely actualising all potency to move? Regents do not address this problem, even if we can formulate an answer on the basis of their philosophy. I suppose that the idea that no act whatsoever can completely satisfy the potency of natural bodies is a stronger principle, and that natural rest must be interpreted within the philosophical framework of act and potency. It is thus conceivable that bodies in their natural place retain potency towards movement, because a complete actualisation of their potency would bring about that bodies are not what they are: they would be a different type of compound.

3. The movement of *gravia* and *levia*

Regents dedicate most attention to the analysis of the third point concerning heavy and light bodies: how they move, and what the mover is. The regents' century was closed by the grand Newtonian picture of a universe structured and held together by the law of gravity, an epoch-making revolution, which heavily influenced teaching in Scottish universities. Until 1650 we still find a predominant Scholastic view, which surfaces from time to time up to the 1720s. In a graduation thesis by Anderson 1720, XIX, we read the following words:

> *Scholasticorum* Commenta de *Fuga Vacui et Levitate* Corporum *absoluta*, certissimis experimentis, eliminata sunt; quippe demonstratum est ipsum Aerem, aliaque omnia Corpora Terram ambientem versus ejus centrum gravitare; ea vero, quae Levia dici solent, sursum pelli, propterea quod fluido Aeris, cui innatant, minus sunt gravia. Eadem gravitate, tanquam universali Naturae Lege, omnia Systematis mondani Corpora, versus se mutuo urgeri, demonstravit praedictus Eximius Auctor.[9]

[8] I deal with the interesting set of 1626 *Theses* in the concluding part of this chapter.

[9] John Anderson, *Theses philosophicae*, Aberdeen 1720. The *Eximius Auctor* is Isaac Newton.

What I find interesting is the threefold grammatical form that the word *gravitas* takes on, to which three aspects correspond: 1) 'gravitare', line five, in verbal form, denoting an action of bodies; 2) 'minus gravia', line seven, an adjective referring to bodies; 3) 'gravitas', line eight, noun: the Newtonian concept, referring to a physical law. Until 1720 we find evidence of an enduring Scholastic heritage, despite the enthusiastic reception of Newton in Aberdeen, where Anderson was a regent.[10]

In the seventeenth century, the theoretical development regarding the movement of heavy and light bodies saw a shift from movement as directed and caused by an agent, to movement as a natural and inseparable state of bodies.[11] The nature of the mover of heavy and light bodies was widely debated in late Scholasticism and it is one of the doctrines destined to undergo the deepest changes in the following decades. What matters now is the Scholastic antecedent of the Scottish reception of Newton.

Regents usually divide themselves on the nature of the mover, which can be either internal or external. An internal mover is the very form of a substance, say, the form of a man is the mover of the substance man; an external mover is instead something external to the moved substance causing it to move, say a man tossing a stone. On a general level, the former movement is called natural and belongs to things which are self-moved, the latter belongs to inanimate things, and it is called violent (*violentus*). The spectrum of possible movements is not restricted to these two types: regents believe that while we have an absolutely natural movement, we never experience an absolutely violent movement. In fact, whatever a thing can do is somehow permitted by its nature: this way, violent is not to be understood as in contradiction with a body's nature, or negating a body's nature while occurring. In the example of a stone tossed upwards, this movement is violent because a stone never jumps upwards alone, yet it is natural because it is not contradictory that a stone goes upwards when pushed with sufficient strength. Some regents conclude that every natural movement is in the end a mixed movement.[12]

[10] David B. Wilson, *Seeking Nature's Logic*, University Park (PA), Pennsylvania State University Press, 2009, ch. 1.

[11] As in Galileo, or in Descartes, *To Debeaune* 30 April 1639, *AT* II, pp. 543-544.

[12] *"Esse naturalem aut violentum sunt tantum accidentales differentiae motus ex parte principij, a quo non sumitur unitas vel distinctio specifica."* Stevenson 1625, TP XVI.4.

3.1 Generans as external principle of movement

The doctrine that the principle of movement of heavy and light bodies is external is Aristotelian, and, among others, was held by Thomas Aquinas and the Coimbra commentators. Adamson 1600 and Lesley 1625 both refer to book VIII, 4 of Aristotle's *Physics* as a key passage:

> Et sibi, et veritati consentaneus est Philosophus, dum cap. 4 lib. 8. Phys. contendit Gravia et Levia moveri ab externo generante, et impedimentum removente, nec ullum habere internum sui motus principium activum: Cap. autem ultimo, ab internis et propriis formis ea asserit agitari. [Adamson 1600, TP VI]

> An gravia et levia ab externo tantum principio moveantur? Aff. *Arist.* 8. *Phys.* c. 4. [Lesley 1625, Problemata physica 9]

Setting aside the contradiction that Adamson sees between the two Aristotelian accounts, the strength of this theory lies in the distinction between animate and inanimate beings, as Coimbrans claim in their commentary on book VIII, 4, 1-3:

> Haec controversia tribus conclusionibus dirimenda est. Prima sit: gravia et levia, cum in naturalia loca tendunt, non moventur ab se, ut a principe causa sui motus. Haec ita probatur: movere se simpliciter et ut principalem sui motus causam, est proprium munus vitae; atqui elementa non vivunt; nequeunt igitur eo pacto sese movere. [...]
> Sit secunda conclusio: gravia et levia, quoties naturalia loca petunt, moventur a generante ut a principe causa effectrice sui motus. [...] Hoc medium, praeter alia est ipse corporum gravium et levium motus; ergo a generante efficiendus erit eidemque attribuandus. [...]
> Sit tertia conclusio: gravia et levia non habent in se principium passivum duntaxat suorum motuum, sed moventur effective a propria forma, ut a principali instrumento generantis, itemque ab insita gravitate et levitate, ut a minus praecipuo instrumento.

We will see what replies regents have for the Coimbrans' conclusions. The strongest argument in favour of an external principle of movement is conclusion 1: if we accept that the form of heavy and light bodies is the principle of their movement, there seems to be no strong distinction between animate bodies and inanimate bodies. Consequently, the definition of 'nature' would equally apply to animate and inanimate. In conclusion 3 the Coimbrans grant some sort of causal power to the form, but only the causal power typical of instruments: in this case, these are instruments whose power comes from the mover (*generans*). Fairley 1623, TP XIIII, defines instrumental cause as follows:

> Causa principalis, et instrumentalis, quod ad modum operandi, in hoc distinguuntur; quod causa principalis operetur per virtutem propriam, et non ut virtus alterius, instrumentum vero praecise in quantum virtus alterius.

What the Coimbrans have in mind is that the mover (*generans*) sets heavy and light bodies into motion not directly, but by giving them their actual form: *"ideo causa motus ipsorum dicitur esse generans, qui dedit formam."*[13] A cause has in itself all the causal power that is transferred to the effect: thus, the mover is the principal cause of the movement of the effect, even if the effect's form (the form of heavy and light bodies) acts as instrument. An instrumental cause is a true cause, it is simply not the primary cause.[14]

3.2 Form as internal principle of movement

A more successful theory among the regents is that the form of heavy and light bodies is properly called the principle of their movements.[15] Regents offer replies to the position of the Coimbrans concerning the distinction between animate and inanimate and the role of the mover.

Adamson, directly after quoting Aristotle's theory, puts forwards his own:

> Ordine naturae primum movetur Grave (de Levi iudicium idem) a sua forma agente per emanationem: secundo totum Grave suo motu movet medium, ut agens per transmutationem. [1600, TP VII]

The talk of causality is not available anymore, since the form of heavy and light bodies cannot be in the relation of cause and effect towards its own substance: rather, heavy and

[13] Thomas Aquinas, *CG*, l. 3, c. 67, n. 2.

[14] Schewer 1614, TP XXIV agrees with the Coimbrans, and his thesis closely resembles *In Phys.* VIII, 4, 1-3.

[15] On this view, see also: King 1620, TP XIII.3; Wemys 1631, TP XV; Leech 1638, TP 30.

light bodies are moved 'by emanation', a relation which can occur between a form and its accident (or, in another context, between the object known and the intelligible species emanating from it, not caused by it). With respect to the sort of relation that is in place in the movement of heavy and light bodies, Adamson's and the Coimbrans' theories are deeply different.

King 1616 broadens the spectrum of the analysis even further:

> Elementa non moventur ab aliquo externo, sed proxime et per se a suis formis, ac motorem internum habent.
>
> Non est necesse, ut quicquid per se movetur constet ex parte movente, et parte mota: sed solum quae perfecte, et per se a se ipsis moventur, cujusmodi sunt animata.
>
> Elementa ab animantibus in hoc distinguuntur, quod haec non solum motus sui principium activum, verum etiam (ut loquuntur) initiativum in se habent, cum a se moveantur, et a se incipiunt moveri: illa vero etsi moveantur a se, nempe a propriis formis, non tamen a se incipiunt moveri, sed ab externo, generante nempe, aut removente impedimentum.
>
> Cum Aristoteles negat elementa a se ipsis moveri, nil aliud vult, quam ea non eo modo moveri quo animantia, quae undecunque, quocunque, et quandocunque volunt seipsa movere possunt. [TP XIII]

We can take this passage by King as the standard reply in the Scottish universities to the Coimbrans. There are a number of aspects to underline: 1) elements (and consequently bodies) do move themselves in virtue of their forms, like an 'inner motor'. 2) The objection can be raised that self-movement contradicts the principle *omne quod movetur ab alio movetur*, and that if forms move heavy and light bodies, a further mover is required for forms: King replies that it is enough to assume the same scheme for animate and inanimate beings. Animate beings are in movement as a whole, in virtue of their form as essential part of the moving whole. 3) The analogy between animate and inanimate does not hold any more when it comes to what King calls the 'initiative' of movement: animate beings can decide when and how to move, while inanimate heavy and light bodies are forced in their rectilinear downwards or upwards movement and cannot decide when to move.[16] It is their nature which enables them to move, yet they need something external to them to move: a mover acting, or the removal of an impediment (*remotio impedimenti*). If a cup is

[16] Reid 1622, TP XX.4: *"Facultas loco motiva non constituit gradum vitae a sensitivo distinctum, in ordine ad principium, sed duntaxat ad subjectum, in quo quandam perfectionem (sed accidentalem) importat."*

on a table, the table is the impediment to the cup falling: this impediment is preventing the cup from following its nature as a heavy body. Were the table removed, the cup would by its nature move downwards. Yet, the removal of the impediment is the cause of movement only by accident. In contrast, a cat sitting on the table always has it within its powers to move down from the table:[17] in normal conditions, a cat does not require the intervention of an external factor.

Regents and Coimbrans do agree on one aspect: *gravitas* and *levitas* are natural powers following from the essence of bodies. What they disagree about is how determinant these powers are in causing movement.

Sibbald, regent at Marischal College, wrote in 1626 a set of theses almost ad hominem against Thomas Aquinas. There he rejects Thomism on the distinction between existence and essence, on resolution into prime matter, on the principle *unius generatio est alterius corruptio*, and on the role of *generans* in the movement of heavy and light bodies. The passage on forms is interesting in the rejection of a Thomistic doctrine that we have seen accepted by the Coimbrans:

> Neque gravitas et levitas proprie dici possunt generantis instrumenta, sed geniti, cum ab ejus forma emanent, ab eadem conserventur, et ab illa tanquam principali causa immediate agendum applicentur quae tamen in generantem minime quadrant, quae tantum dedit necessaria ad finem (formam producendo) virtualiter et in radice, non formaliter et in se, ut loquuntur. [1626, *A quo moveantur gravia et levia*, 2]

The key remark is that *gravitas* and *levitas* cannot be called instruments of something external to their form (the *generans*), because they inhere as powers in their form, on which they depend. The only dependence Sibbald acknowledges is the dependence of the substance on the *generans*, which causes a substance to exert all its movements on its own. Being heavy and being light are thus instruments of the form of their substance: in other words, a heavy or light substance does move itself.

[17] In this case the movement of the cat would not be rectilinear. The example holds because the stress here is on where the 'initiative' of the movement is: within or outside the moving body.

3.3 Form as nature, nature as *finis*

The analysis of the regents' theories of movement in general and local movement in particular allows us to understand the context of the doctrine of the identity between form and nature. Aristotle in *Phys.* II, 1, 193 a 27 ff. reaches two conclusions: nature can be intended both as matter and as form. Matter is the subject of all substances, and hence something all substances are from: and this is one meaning of nature. Yet, the prevailing meaning is nature as form, because the thing all beings aim at is more important than the thing all beings come from. Thus, form is nature, and nature is the end of beings.

Regents often comment on this theory, endorsing it. It is a famous and non-controversial Aristotelian passage, which in turn does not raise a debate in the *Theses*.[18] Yet, the theory of form as nature is required to complete the account of movement. We have seen that a significant majority of regents holds that heavy and light bodies are moved by their own forms, which are the nature of the substances they inform, and which are also the end of their substances. The identity form-nature-end is expressed throughout the *Theses*. When a body is in movement, its aim is the acquisition of a new form (*terminus ad quem*): alongside the formal distinction between movement (flux of form) and its terminus, we can affirm the identity between form and terminus once the movement is complete. Rest is the acquired acquisition of a new form, so it is ultimately more perfect than movement.[19]

One important aspect is that the definition of nature as 'internal principle of movement' must make room for the inclusion of passivity. The result is that nature is not only an '*active* principle of movement' but both an '*active* and *passive* principle of movement'. The case of the movement of heavy and light bodies makes this point clear. A further application of this theory is evident in the analysis of the movement of the heavens: their

[18] For the most explicit passages on form as nature: Robertson 1596, TP 10; Carr 1617, TP V.5; Forbes 1623, TP VI; Rankine 1627, TP VI.2. The theory of form as nature inscribes itself in the general picture of the natural philosophy of the theses: in fact, form is the end of matter, nature is the end of the compound, and form is what gives the essence to the compound: therefore, form is nature. Also, form can be interpreted as the mover of inanimate bodies, and nature is the principle of movement: therefore form is nature. The theory of form as nature thus surfaces in all the regents who hold any one of these theories.

[19] The question can be asked whether local movement can be included in this picture. In fact, as we are reminded by Reid 1614, TP 6: *"motus localis a caeteris distinguitur, quod terminus ipsius quod externum sit, aut saltem respectum ad extrinsecum includat."* The theory that local movement enjoys characteristics on its own is present in a minority of graduation theses, but the objection to the general view that local movement is not of a different type from the others still holds. The end of local movement is a whereness, which is not something 'of' a substance (like, say, quality) rather something external to a substance. If this objection is to be brought to its extremes, 'whereness' would not be a category any more, but a relational position in space. On the contrary, if we still consider 'whereness' as a categorial predication, we can still say that an *ubi* is predicated 'of' a substance.

movement is said to be natural even if it is originated by the external intelligence (*intelligentia*).[20]

Regents argue that the movement of heavy and light bodies is natural independently of whether they consider the nature of heavy and light bodies as an active or passive principle of movement: thus, the stress is on 'natural', more than on 'active or passive'.[21] Now, some regents hold that the principle of such movement is to be found in an external agent, with the forms playing the role of instrumental causes: also in this case, this movement is natural. Thus, we must expand the definition of nature as Craig 1599 does in TP 6.1:

> Motus etiam ille, qui ab externo est agente, cuius passivum principium est internum, naturalis est dicendus.

The reason for the need for expansion is that in the presence of an internal passive principle, the conditions for a violent movement per se are not met. The body in movement in this case is naturally open to receiving the determination towards this particular movement, so that the agent causing the movement does not coerce the nature of the moved. It is simply the case that the moved body alone does not have the power to bring about such movement. In conclusion, nature can include both an active and a passive principle of movement.[22]

Scholastic natural philosophy exploits the notion of finality at many levels: from the individual directedness of the movement towards its terminus, to the general directedness of matter towards form, up to the universal directedness of the universe towards perfection, and ultimately towards God. The regents perceive the intrinsic finality of creation as something more than just a successful explanatory theory. In King 1612 we find a reference to the behaviour of the wise man whose echoes extend as far as the eighteenth century, in the words of George Turnbull, regent at Marischal College and teacher of Thomas Reid:

> Non sunt multiplicanda entia sine necessitate: debet enim sapiens naturam imitari quae nihil frustra facit. [TP 6]

---

[20] See below, part II, chapter 3.

[21] For example: Young 1613, TP 1.VI; Forbes 1623, TP VI; Rankine 1627, TP VI; King 1628, TP V.1.

[22] Reid 1622, TP XI.7, mentions the case of blood, whose movement is by an external principle (vital spirits) and also natural.

> Omnino fatendum est mundi corporei ordinem elegantissimum maximeque concinnum esse. Illoque certe nobis optimum vitae et morum exhibetur exemplar.[23]

With respect to generation, the directedness of natural processes includes any individual being, animate or inanimate, which is brought about in order to exert some operation (Stevenson 1625, TP VI): in nature, no being is produced without an end, and the totality of beings is one per se, not merely by accident. The totality is unified, for instance, by the universal end of sustaining life (human life above others), which, we will see in the next chapter, is the end of the heavens. Graduation theses are one example, among many others, of the interpretation of the Aristotelian doctrine in *Phys.* II, 8 of the finality of nature in a Christian philosophy.

One theoretical aspect of Scholastic natural philosophy is the endorsement of final causality. We find some detailed passages in many theses in which the regents express their view on a subject which, by the time the earliest theses were written, was under attack by the so-called *Moderni*. In fact, final causality has been generally rejected outside of Scholasticism as a consequence of its being taken as an anthropocentric approach, in conflict with the new science.[24] It is thus interesting to see what regents believe final causality to be. Once again King is one of our main sources:

> Τέλος ἐσιν τὸ οὗ ἕνεκα. [*sic*] 2. *Post.* 11.
> Finis igitur non est causa, nec habet rationem finis prout actu jam agenti adest ab eo acquisitus.
> Quumque finis sit qui explet appetitum agentis, quo praesente cessat actio, et in cujus possessione

[23] G. Turnbull, *Theses philosophicae de scientiae naturalis cum philosophia morali coniunctione*, Aberdeen 1723. It is arguable that similarities do not stop here: despite the stress we find in King on the fallibility of human will and intellect due to the original sin, both regents share the confidence in philosophy as *"medicina morborum animi"* [King 1612, TP 1]. I believe that these words are not a novelty per se; still, the continuity in Scottish universities of these themes over more than a century and amidst great changes in philosophy is remarkable.

[24] Final causality shapes Scholastic natural philosophy as a whole: from the form-matter structure of the compound, to the nature of the celestial movements. It seems that the modern philosophical reaction (especially Descartes') to final causality in natural philosophy focused primarily on the movement of inanimate bodies and of the non-rational animate bodies; and secondarily on all the other occurrences of finality in Scholastic philosophy - with, of course, the exception of rational beings and intentional finality. The graduation theses underline the difference between efficient and final causality in the light of ruling out the theory that final causes act as physical causes. Wemys 1631, TP V pursues a different strategy, similar to Buridan's (Des Chene, *Physiologia*, p. 187): granted that the final cause is active only when apprehended by a mind, Wemys claims that: *"Finis non influit in effectum, nisi mediante efficiente."* This is an attempt to understand finality in terms of efficient causes, without holding that they are the same kind of causality. Yet, this is the only case in which final causality can be "downplayed" without endangering the general structure of Scholastic philosophy: in fact, as an example, the concepts of appetite, good and form as the end of matter are the foundation of the very notion of substance.

> agens conquiescit, non erit causa secundum esse intentionale quod habet in mente agentis.
>
> Quare finis proprie et per se causat, secundum esse reale extra animam futurum et possibile acquiri ab agente.
>
> Nihil igitur impedit quo minus non ens actu, quod tamen esse et a nobis parari potest, etsi effective causare non potest, nec Physice movere, causet tamen finaliter movendo agens motione quadam metaphorica. [1612, TP 8]

> Finis vere impossibilis, apprehensus tanquam possibilis; movet voluntatem ad veros actus reales et physicos.
>
> Ad essentiam causalitatis finis, sufficit bonitas realis apprehensa, licet ad terminationem requiritur vera. [...] Motio finis ejusque causalitas, non est intelligenda ad modum causarum modo materiali causantium. [1628, TP VI]

In these dense passages King accepts the validity of final causality in natural philosophy by offering an account of its essence. In fact, final causality is different from material, formal and efficient causality, since it does not act in the way these natural causes act: ergo, it is not a natural cause. Yet, there is still a role for it: final causality requires the mediation of a mind which apprehends the good of an aim and consequently brings about physical actions in order to acquire this good. An end is always (whether it is per se or because it is thought to be such) a perceived good. This 'being apprehended' by the agent suffices to have a final causality, since the agent acts in order to acquire this good. To complete the acquisition though, an apprehension of the good is not enough because the acquisition of the good must be real and physical.

This is the account we find in King, mainly based on the example of a mind perceiving a good, and prompting the agent to move accordingly. Needless to say, final causality is more problematic if there is no mind. Descartes' famous objection was exactly that Scholastic inanimate bodies would resemble rational ones by actively aiming at an end.[25] The only acceptable notion of end should be an end perceived as such by a mind within a natural process which per se does not entail finality. Finality would then be reduced to causal efficiency.

And indeed late Scholasticism was not far from this account of natural finality. If we also consider the position of Suárez, the Coimbrans, Fonseca and Goclenius we realize that the 'intentional being' of an end is considered a conditio sine qua non of the causality of an

[25] In a letter to Mersenne in 1643 (*AT* III, p. 648) Descartes expounds his reading of Scholastic real qualities as *"petites âmes à leur corps"*, which entails the notion of anthropomorphism.

end: an end does not cause by its intentional being though, rather the good of the end sets will in motion in order to acquire it.[26] Between a perceived end and will there is not a relation of cause and effect, unless we intend it 'metaphorically'. The shift from final to efficient causality is realized when the agent acts to acquire the end.

When regents speak of final causality they accept this general framework of mind-perceived good (*esse intentionale*), and in no cases has final causality a place in non-rational beings. On a universal scale, thus including inanimate bodies, the same structure holds: we have seen that heavy and light bodies act in virtue of their forms, which are their natures. These natures are given by God in the act of creation. We can appreciate now one of the strongest reasons in favour of the theory of *generans* as principle cause of movement: heavy and light bodies do act in such and such a way because they are given such and such a nature by God: this also explains the actions of inanimate beings according to final causes.

Despite being mutually opposed in respect of what the principle of movement is, the *generans* theory and the form theory entail a deeper agreement on the nature of final causality and movement in general.

3.3.1 An exception? Strachan 1631 on *medium demonstrationis* and *intentio metaphorica*

Andrew Strachan, regent at King's College between the late 1620s and early 1630s,[27] deals with natural finality in a complex passage, both on a theoretical and a grammatical level. According to Strachan, Aristotle's original doctrine has it that the heavy and light bodies are the intrinsic causes of their movements[28] and Thomas and Scotus corrupted Aristotle on this matter [TP IV]. I now quote the first part of his own theory:

> Nihil proficiunt, qui demonstrare laborant gravium et levium naturam, esse causam principalem motus ipsorum: argumento petito a natura demonstrationis: quia viz. per naturam eorum demonstrari potest

[26] D. Des Chene, *Physiologia*, pp. 186-200.

[27] We have little information regarding Andrew Strachan: regent for the graduates of 1628-29, 1629-30, 1630-31, 1631-32 at King's College, Aberdeen. His only extant theses are 1629 and 1631. Later on Professor of Divinity. *OG*, p. 55.

[28] Strachan quotes *De Coelo* I, 4, and offers a reading of Aristotle incompatible with that of Lesley 1625 and Adamson 1600 [above, section 3.1], who quote *Physics* VIII, 4. In TP IV.4, Strachan claims that Aristotle interpreted the *generans* as principal cause of movement only with regards to causes operating by emanation, thus not absolutely speaking.

> ipsorum motus: in omni autem demonstratione potissima medium debet esse causa principalis: absque qua foret ut non ingeneraret perfectam scientiam. Non enim motus sed mobilitas demonstratione potissima concludi potest de gravibus et levibus per ipsorum naturam. [1631, TP IV.1]

The key passage is that the theory that nature is the principal cause of the movement of heavy and light bodies is proved false by the very nature of demonstration: there is no demonstration starting from the nature of heavy and light bodies which proves their movement. Why is this so? In a scientific demonstration (*potissima demonstratione*), the middle term must be the principal cause. The conclusion about heavy and light bodies on the ground of their nature thus can be 'mobility' (*mobilitas*), not movement.

More elements are required to understand what Strachan has in mind. In a demonstration delivering perfect science, that is a universal and necessary conclusion, the role of the middle term of the demonstration (*medium demonstrationis*) is unique and universal; it must convey a proper knowledge of the thing to be demonstrated, and somehow the conclusion is posited as soon as the middle term is posited. If these conditions are not respected, then no conclusion can be reached. Strachan holds that the nature of heavy and light bodies cannot play the role of middle term, because, and this is his claim, the nature of heavy and light bodies only lets us conclude about the mobility of these bodies, and not about the type of movement they undergo. In addition, some sort of finality is required, which specifies the mobility as 'movement downwards or upwards towards a natural place'.

In the same year 1631 a set of graduation theses by Wemys addresses the same subject as follows:

> Medium in demonstratione διοτι [*sic*] est principalis causa affectionis demonstratae.
>
> Forma ergo gravium et levium sunt principales causae eorum motus. Idem est movens et mobile potestate et actu. [TP XV]

According to Wemys, forms are the principal causes of movement and can be the middle of a scientific demonstration of movement. The two regents do agree on the structure of the perfect demonstration: they differ on what can be accepted as the middle of such a demonstration. Wemys speaks of 'form', Strachan of 'nature' and perhaps the disagreement lies in this terminology. In fact, Strachan seems to hold that form alone is not the nature of bodies and that matter must be included as well. Given that matter is essentially open to any movement because matter is the unique subject of all natural bodies, it is then coherent to say that the nature (form and matter) of heavy and light bodies just allows us to infer their mobility, and not the kind of their movement.

The second part of Strachan's passage is the most complex, and touches the notion of metaphorical intention:

> Intentio metaphorica (quam generantia inanimata alunt, ad perficiendum omnibus numeris ea quae ab ipsis generantur, quod tum demum praestant quando illis in loco naturali contingit esse, quem per motum consequuntur) non magis abjudicat naturis gravium et levium rationem causae principalis: quam intentio animatorum quae formalis est (qua in generatione proponunt sibimet conferre genitis a se ea omnia quae ipsorum naturae debentur) aut intentio causae universalis et primae (qua omnium entium perfectionem intendit per media ipsorum naturis consentanea) ponit, aut probat generans animatum aut primam causam esse causas principales, et proximas earum actionum quae a genitis animatis, aut causis secundis producuntur. [*ivi*, TP IV.2]

It might be useful to quote the passage without the parts in parentheses: *"Intentio metaphorica [...] non magis abjudicat naturis gravium et levium rationem causae principalis: quam intentio animatorum quae formalis est [...] aut intentio causae universalis et primae [...] ponit, aut probat generans animatum aut primam causam esse causas principales, et proximas earum actionum quae a genitis animatis, aut causis secundis producuntur."*

Strachan's point is that just as the apprehension of a good as an end does not necessitate our will to pursue the end, so the 'metaphorical intention' does not deprive the natures of heavy and light bodies of their being principal causes. The *generans* of animate bodies is the first cause (God), yet we do not say that it is the cause of animate bodies' actions: the same holds for heavy and light bodies.

Metaphorical intention is a hapax legomenon in the *Theses* just as Strachan's theory is. In a bracketed line, the regent tells us that metaphorical intention is given to inanimate bodies by generators (*quam generantia inanimata alunt*): this would mean that finality is within the inanimate body, not because the first cause has externally intended it to act in a finalised way; rather, inanimate bodies are intrinsically finalised. And also, the notion of 'metaphor', usually employed in a mind-object context, is accepted by Strachan with regards to inanimate (ergo mindless) bodies.

4. Reid 1626[29]

Reid 1626 is the last set of graduation theses by this regent, who taught in Edinburgh university from 1606 (the beginning of his first four-year curriculum) to 1626. It is a very important work to understand some of the changes which were occurring in the philosophy of the regents. The most interesting field is movement, but other features are remarkable as well. For example, a change in the format of graduation theses is clear if we compare 1614 with 1626: The 1614 *Theses* are written in the form of a commentary on the *Physics* of Aristotle, each thesis usually consisting of a quote of Aristotle that the regent is commenting on with a number of corollaries and notes. 1626 instead is structured more as a little treatise, still very much focused on the *Physics*, yet with a unity more thematical than expository. Interestingly as well, the only two authorities mentioned in 1626 are Aristotle and Scaliger, which gives us an idea of what philosophers Reid draws inspiration from.[30] The combination of 'the Philosopher' and one of the most recent Renaissance philosophers is not rare in the *Theses*, seen when discussing Transubstantiation.

I think that Reid 1626 is still fully within the Scholastic tradition. The regent does not reject Scholastic natural philosophy in its key aspects, as he accepts the analysis of prime matter, of heavy and light bodies, of natural places and the subordination of philosophy to theology in those subjects in which a conflict is possible. Nonetheless, Reid's theory of movement shows some unique features which are at odds with the work of the other regents. The central feature seems to be a different account of the relationship between movement and rest, and consequently nature and rest.

Here are his words on the subject:

[29] The translation of the *Theses physicae* of Reid 1626 is in the Appendix.

[30] As we can also understand from the translation of Reid's physical these in the Appendix, the strategy adopted is to make use of Scaliger in the interpretation of Aristotle: it does not seem to be the case that Reid sees Scaliger as in opposition to Aristotle. I believe that the same attitude is present in all the references to Scaliger in the *Theses*. In general, Scaliger is the most quoted non-Scholastic Renaissance philosopher. The regents usually quote Renaissance Scholastics: Suárez, Zabarella, Gabriel Biel, the Coimbrans, Ruvius, and Cajetanus are the most quoted. The favourite non-Scholastic Renaissance sources are Scaliger and Ramus, even if the latter has a very minor impact on the regents' natural philosophy. In general, the regents' sources are still much in the style of medieval Scholasticism: Thomas, Scotus, Durandus, Albertus Magnus, Averroes, Avicenna, Plato, Augustinus, the Nominales, Porphyry, the Greek physicists. Yet, the post-Humanist character of the *Theses* is clear in the constant references to the Greek commentaries of Aristotle and to Classical Latin authors: in particular Alexander of Aphrodisia, Simplicius, Plyny, Seneca, and Cicero. The overall picture seems to be one of continuity with medieval Scholasticism in terms of references and debates; nonetheless, Scaliger is the most apparent example of the assimilation in the Scholastic philosophy of the regents of some innovations of Humanism.

> Terra non majorem habet propensionem ad quietem, in infimo loco, quam ignis ad motum, in supremo. Natura terrae tantum motus, non quietis ipsius, principium et causa est. Et terra tantum mobilis, non immobilis est sua natura, ac conditione naturali.
>
> Cum itaque omnis quies, etiam qui motum naturalem sequitur, ejusdem motus privatio sit: natura ut quietem proprie non expetit, ita nec eandem intendit.
>
> Unde inferimus primo, longe differre, unumquodque in suo loco naturali manere, et in eodem quiescere: nam illud omni corpori naturali naturale est, hoc nulli corpori naturali naturale est.
>
> Inferimus secundo, naturam nihilominus, etiam principium et motus et quietis dici copulative, si quies fundamentaliter non formaliter, hoc est pro ipsa possessione ac fruitione formae ac termini, non simpliciter pro motus privatione accipiatur. Atque hoc sensu, idem est κινεῖσθαι καὶ ἵσασθαι, et idem est κινεῖσθαι καὶ ἠρεμίζεσθαι ex Arist. 6. Phys. 8. text. 67. [TP II]

These extracts are Reid's comments on *De Coelo* I, 3, quoted as the heading of the thesis, where Aristotle writes that heavy bodies are those which are underneath and go downwards, while light bodies are above and go upwards. A traditional Aristotelian doctrine, whose comments lead us away from it.

In the first lines, Reid claims that rest for earth is equivalent to movement for fire. This theory is not new to the *Theses*: Lesley 1625, TP XIV, has it that fire *"movetur ut moveatur, non ut quiescat"*, and quotes Zabarella in support of this view. As we will see in the next chapter, the movement of heavens is also thought to be an essential condition, not a movement towards a greater perfection. In sum, regents speak of movement as an end in itself in specific contexts. What Reid does differently, is predicating movement as proper to fire in parallel with rest proper to earth, as if an analogy of proportion 'rest : earth = movement : fire' were available here.

The following lines clarify the point: the nature of earth is the principle and cause only of movement, not of rest: earth is, by its nature and in a natural condition, mobile. To make things more explicit, Reid openly claims that 'nature by itself does not strive for rest.' I believe we are allowed to see in these words a rejection of the traditional doctrine of the directedness of natural movements: if rest is not the end of movement but just a privation of movement, then what is natural to bodies is not the movement towards an end but precisely movement as movement. A Scholastic could agree that the natural condition of bodies is movement, but would not give up on the idea of rest as the end of movement, rather than simply a temporary 'suspension' of movement.

The first inference from this theory reminds us of an earlier point made by Reid, in 1614, TP 3.7 (section 2.1), where he assumes that 'remaining in place' is the same as 'resting'. Now it is made clear that this is not the case: remaining in place (*manere*) is natural to bodies, while resting (*quiescere*, which implies a full realisation of the nature of bodies, not simply an actualisation of potency) is never a natural state of bodies. The second inference is that nature can be called a principle and cause of both movement and rest only if rest is understood fundamentally as possession of the form, and not formally as privation of movement. In this sense, Reid explains, it is the same 'to move and stop' and 'to move and to rest'. If we look at the form of movement, stopping is no different from resting; if we look at the matter of movement, in this case nature is also the principle and cause of a movement-towards-form (and natural places retain their importance).

Despite a general adherence to Scholastic natural philosophy, Reid brings forward some considerations innovative in the context of graduation theses. I believe that his case is not dissimilar to Dalrymple 1646. It is a matter of speculation what Reid's possible sources of inspiration are. His reference to Aristotle's text does not help much: *Physics* VI, 8 is in fact about the analysis of moving and stopping in relation to the instant, not in relation to the natural places: Reid's theory does not seem to be Aristotelian. I will address the question of the role of Aristotle later in chapter 3 and in the Conclusions. What so far appears to be the case is that regents looked back at the Greek texts of Aristotle as still the most relevant and inspiring works in philosophy.

5. Conclusion

The notions of the heaviness and the lightness of bodies are central to Scholastic natural philosophy. A body moves according to its nature: heavy and light bodies move according to, respectively, their natures as heavy and light bodies. A heavy body moves towards the centre of the universe, while a light body moves towards the upper limit of the sublunar sphere, limited by the sphere of the moon, the first celestial sphere.

The regents are not committed to an isomorphic concept of space: in fact, bodies tend towards their natural places, where their movement naturally reaches its end, and where the substance reaches a state of rest. This is the most general notion of an 'end' of movement.

Since movement is also the acquisition of form, each time that a new form is acquired, a determined movement ends and rest is reached. In this particular sense, the new acquired form is the end of a determined movement. Nature, which, according to the regents, is the same as form, acts as the final cause of movement. Form/nature is fully realised in its natural place.

The regents answer the question of what is the primary cause of the movement of heavy and light bodies: in response to the Scholastic tradition according to which the generans moves heavy and light bodies, the regents reply that heavy and light bodies move themselves, even if not in the same way as animate bodies move.

We have seen that the regents do not understand the final cause as acting as a physical cause: yet, they still find place for the natural directedness of movements in their natural philosophy.

Reid 1626 seems to put forward a theory of the finality of the movement of fire which breaks with the Scholastic tradition: in fact, the regent claims that fire does not move towards rest, but rather moves in order to move. The end of fire is movement, ergo the rest of fire is movement as well. Reid seems to hold that rest is not a natural state of bodies, and that it is nothing different from privation of movement.

Part II, chapter 3

The movement of the heavens

In the seventeenth century the understanding of the movement of the heavens saw dramatic developments in both epistemology and metaphysics. In the former field, the increasing use of mathematics progressively drove the enquiry on celestial bodies away from a purely philosophical reading; in the latter, the traditional framework of the distinction in nature between sublunar and celestial world gave way to a unitary analysis based on common laws and properties. In a broad scheme, the shift took place from traditional Scholastic accounts to the first forms of modern science. The analysis of the movement of the heavens is one of the most apparent elements of the so-called 'scientific revolution', the great scientific paradigm-shift which paved the way to what modern science is.

This phenomenon falls within our scope since we have to investigate what graduation theses say with regards to cosmology. Such an investigation will enable us to establish the extent of the Scholastic influence and the extent of the possible early penetration of the new science in Scottish universities up to 1650. Scottish Scholastic natural philosophy on this matter is heavily indebted to Scholasticism, as was much of the European philosophy as a whole. Indeed, even before the scientific revolution, the Scholastic approach was not the only available approach to cosmology, as Renaissance philosophies developed alternative ways to give answers and raise problems about the nature and movement of the heavens. But if Scholasticism was not the only system available, it was certainly the most widespread, inclusive and influential.

In Scottish universities in particular there is no evidence of the acceptance of philosophies other than Scholasticism, even if philosophers such as Pico della Mirandola or Giordano Bruno were read and studied: the background is then Scholastic. What of the outcome? I shall argue that graduation theses show examples of proximity to some theories of the *Moderni* while still being deeply rooted in the Scholastic tradition. I do not intend to read graduation theses in parallel with contemporary scientific works: this approach would find little textual evidence in the graduation theses and the very choice of contemporary authorities would inevitably be arbitrary.[1] What I set out to do is to offer an account of cosmology in Scottish universities against its Scholastic background. The voice of modern science, though not absent, is barely audible. This is, I believe, an interesting historiographical point: Scholasticism, in all its confessional, national, school-based forms offered so many solutions and alternatives that, for instance, the Scottish regents could accept the notion of a void and still be Scholastic. If we want to draw a parallel with Descartes, can we say that regents were more 'scientific' or more 'modern' than Descartes on this matter? Clearly not; yet, the rejection of the theory of void is arguably one of the most evident 'Aristotelian' elements in Descartes' philosophy.

In my opinion, the historiographical category of the 'old' Scholasticism facing the 'new' philosophy must be dropped if by 'old' and 'new' we mean anything more than chronological succession. In the beginning of the seventeenth century, various philosophies were confronting each other from different if not totally opposed standpoints, nonetheless some theories were in fact shared, and same conclusions reached from different premises. Going back to the example of void: does the rejection of a void make a philosopher Aristotelian, and vice versa? The answer is again 'clearly not'. If we limit Scholasticism to either a narrow or broad set of doctrines that a philosopher must commit to in order to be 'Scholastic' (and the same can be stated about Aristotelianism), we risk losing sight of the historical variety of Scholasticism in favour of a merely philosophical and historiographical unity.[2]

The movement of the heavens is a form of local movement. In the Scholastic theory of movement, local movement is the type of natural change occurring when a substance acquires a new *ubi*, a new presence in space. The heavens were traditionally intended to be immutable, which means not subject to generation and corruption, thus not subject to any movement which implies the corruption of an old form and the acquisition of a new one. Ergo, the heavens are not directed towards an end, since the end of natural movement is the new form. Local movement of the heavens is of a type of movement which does not include directedness. This chapter is then about 1) the nature of the heavens; 2) their movement, with particular attention to the theory of void; and finally 3) the extrinsic finality of their movement. One last point is about the reception of Aristotle's proof of the

---

[1] What I mean is that the new or modern science was itself a vast spectrum of sometimes mutually divergent and incoherent theories, not a unitary body. Why then prefer, for example, Galileo to Descartes, or vice versa, in absence of historical evidence in the graduation theses?

[2] This seems to be the approach of the otherwise valuable introduction to a standard version of Scholasticism in W. Ott, *Causation and Laws of Nature*, ch. 3. The author seeks to sketch the most widely held positions by an almost exclusive reference to Thomas Aquinas and Suárez. If this approach can be theoretically fruitful, it is nonetheless historically reductive.

existence of the prime motor, which is the archetype of the Christian demonstration of the existence of God by its effects (*per effectus*), otherwise called 'a posteriori'. Some regents reject this demonstration and I will argue that they do so also on the basis of their confession. The analysis of the reception of Aristotle will continue in the Conclusions.

1. Nature of the heavens

The doctrine according to which the heavens are of a different nature than the sublunar world is the product of a number of theories, assumptions, and arguments all concurring in the same conclusion. It is an example of what we might call a paradigm (scientific, philosophical, cultural) proper to Scholasticism and Aristotelianism, or better, deeply coherent with the historical forms of Scholasticism and Aristotelianism. It is a theory which shaped the cultural world for many centuries in Europe, to be fully rejected only during the seventeenth century. Indeed, regents still subscribe to it in large numbers.

Perhaps more than other theories, this doctrine illustrates the idea of a "paradigm" applied to the history of philosophy, and consequently, to philosophy.[3] Scholastics employed a variety of arguments to prove this doctrine, arguments whose form is based on a number of assumptions and other theories proper to Scholasticism itself and derived from Aristotelianism. Outside this context such arguments are ineffective, if contrasted with many other Scholastic arguments, which may retain their validity. On a deeper level, it is also arguable that such a doctrine is never proved in a satisfactory way: for the reason that, in Scholastic natural philosophy, this doctrine sometimes works as a conclusion, and some times as a premise, and, more importantly, for the reason that every argument within the same paradigm always confirms the paradigm, either directly or indirectly.[4]

---

[3] For the reception of T. Kuhn's paradigm theory in the humanities, *Paradigms and Revolutions*, edited by G. Gutting, Notre Dame (IN), Notre Dame University Press, 1980.

[4] In *The Structure of Scientific Revolutions*, Chicago - London, University of Chicago Press, 1996, Kuhn claims that the achievement of classics of science, such as Aristotle's *Physics*, *"was sufficiently unprecedented to attract an enduring group of adherents away from competing modes of scientific activity. Simultaneously, it was sufficiently open-ended to leave all sorts of problems for the redefined group of practitioners to resolve. Achievements that share these two charcteristics I shall henceforth refer to as 'paradigms', a term that relates closely to 'normal science'"* (p. 10). In the practice of 'normal science', the paradigm sets the nature and direction of research, and *"when the individual scientist can take a paradigm for granted, he need no longer, in his major works, attempt to build his field anew, starting from first principles and justifying the use of each concept introduced"* (pp. 19-20). In some sense, this picture applies to philosophy as well: for example, the difference in nature between sublunar and celestial world shares the characteristics of a paradigm, including the resistance against the paradigm-shift in the direction of modern science.

This doctrine shaped philosophy so profoundly that no Scholastic until the Renaissance really doubted it. It was a paradigmatic doctrine, and only a 'revolution' in philosophy could bring about a true contender to it. Given this picture, natural philosophy was not the only discipline involved: in different ways, moral philosophy and theology benefited from the idea of the universe and man's place in it that would be derived from this paradigmatic doctrine. For instance, in astronomy the geocentric theory was hardly doubted: theologians and philosophers interpreted this scientific evidence of the earth as the centre of the universe to strengthen the Christian idea of the creation made for the advantage of mankind, in a universe ordered by a benevolent maker.

This doctrine and the following scientific revolution have been used by many as a case-study for the shift which occurred in the western world. My intent is much more limited: I intend to show what the *Theses philosophicae* say about the heavens, the form of the arguments employed and how deeply this doctrine is rooted within Scholasticism. But also, how within Scholasticism itself arguments were available for the theory of the identity of celestial and sublunar matter, movement and consequently nature.

The structure of this chapter could have followed the reverse order, with the movement of the heavens dealt with before the analysis of the nature of heavens: this is, indeed, a logical order of exposition, if we accept that the movement of a body tells us about the nature of the body. Or, in a regent's words:

> motus adeo cum natura est complicatus, ut quicquid facit per illum faciat, per illum etiam se nobis patefaciat. [Knox 1605, TP 3]

If when the heavens move they manifest characteristics specific to them, then also the nature of the heavens must be only specific to them.

I shall follow a different order, one which is more *secundum naturam*, for two reasons: 1) the nature of the heavens is logically and metaphysically prior to our knowledge of their movement. Once the nature of such movement is grasped, what is prior according to us (that is, in the order of knowing) must give way to what is prior according to nature (in the order of being); 2) this way of reasoning is Scholastic. I believe this point to be central. Graduation theses are a product of a long tradition, which stretches back to the Middle Ages. The complexity, and wide range of influences (Aristotelianism and Christian revelation above all) in Scholastic philosophy did not allow for a systematic method of discovery in philosophy, much praised and sought after by the *Moderni*. Scholasticism has always been an inclusive way of philosophizing, an exposition of truths either obtained in other disciplines such as theology or possessed for so long that no new proof for them was required. This does not entail that no philosophical progress was ever made: on the contrary, Scholasticism is rich in debate. Yet, philosophy was not about discovering, but about expounding in a more and more inclusive and coherent way.

Putting the nature of heavens first in the order of exposition enables us to make a point about the Scholastic way of philosophizing, and to underline the paradigmatic role of this doctrine.[5]

1.1 Heavens different in nature from the sublunar world

Regents disagree on whether the nature of the heavens differs from that of the sublunar world. The majority says that the two natures are different. The doctrine of the difference in nature is more traditional and more strongly rooted in the works of Aristotle, who dedicated two distinct and complementary works to the sublunar world (*Physics*) and to the heavens (*On the heavens*).

We have seen that sublunar bodies are compounds of form and matter and the subject of all such bodies is the same prime matter. Prime matter thus confers some sort of identity on all sublunar bodies, due to the identity of one of the two principles: prime matter is in fact of the same species in all bodies.[6] Now, 'difference in nature' between sublunar bodies and celestial bodies can mean either one of these two options: 1) celestial bodies are not composed of matter and form, and are not compounds at all; 2) the matter of celestial bodies is a different matter from sublunar bodies.[7] The first option was Averroes's solution, unanimously rejected by the regents: according to Averroes, celestial bodies are pure forms devoid of any matter, hence the difference from sublunar bodies. When regents state this difference, they always conclude that the difference is due to matter; namely that celestial matter is not made of the four elements, and consequently that it is not subject to upwards and downwards movements, which means not subject to finality. In brief, the compounds of form and matter are also to be found in the heavens.

---

[5] The Scholastic way of philosophizing is heavily influenced by the reception of Aristotle's *Posterior analytics*. Yet, in the light of the paradigm theory, it might be asked whether the Scholastics failed to respect the Aristotelian principle that the premises and the conclusions of an argument cannot be interchangeable.

[6] "*Omnis materia sublunarium est ejusdem speciei ex natura sua, utut formae toto genere distinguantur.*" Reid 1622, TP IV.2.

[7] It is important to underline the difference between nature as form (inner active and passive principle of movement) and nature in this context, where nature is taken to signify the structure and essence. The question here is not about the principle of movement of celestial bodies; rather, about what sort of bodies they are.

A body which is not subject to finality is a body which never acquires a form other than its current form. A new form is always the end of movement, it is also said to be its perfection, granted that it is a perfection of something perfectible, not a perfection of something already perfect [which is rest, Reid 1614, TP 3]: this latter case is not available to sublunar bodies, since they are always in movement and unable to fully satisfy their potency. On the contrary, celestial bodies always retain their same forms: celestial compounds are thus necessary, because they are what they can be, and they cannot be any different from what they are. Of course, they can be said to be necessary *secundum quid*, that is, if they are considered from the standpoint of natural philosophy: absolutely speaking only God is necessary.

So, being devoid of potency towards any form different from the current form (no finality) implies that celestial bodies are not corruptible, since the present compound is never going to be dissolved and replaced. Not being corruptible implies not being generable. Celestial bodies are above the natural vicissitudes of generation and corruption, and this is explained by the application of the principle *omnis generatio est alterius corruptio*: if there is no corruption in the heavens, then nothing can ever be generated in the heavens either. Heavens were created by God directly as they are, and were not generated by any created secondary cause.

Fairley 1623 makes an explicit connection between matter and corruption with regards to the heavens:

> Materiae eiusdem speciei habent potentiam passivam essentialem eiusdem rationis, ad easdem formas recipiendas essentialiter ordinatam.
>
> Ergo si materia coeli et sublunarium esset eiusdem speciei eaedem formae continerentur in potentia utriusque, ut forma Solis contineretur in potentia materiae ignis, et viceversa. [...]
>
> Ergo ex eodem posito sequeretur coelestia esse sua natura generabilia, et corruptibilia, quod Arist. repugnat *Lib. I. de Coelo Cap. 3.* [TP VII, 1-4]

Regents seldom speak of fifth essence or quintessence (*quintessentia*), the famous fifth element the heavens were thought to be composed of. Nonetheless, despite this rare use of the word, when they hold the theory of the difference in nature, they implicitly refer to quintessence. We have seen that the four elements are essentially either heavy or light, the property from which movements proper to each one of the elements follow: heavy elements go downwards, light elements upwards. Quintessence is of a different nature, it cannot be said to be either heavy or light: as a consequence, it does not move downwards or upwards. The movement proper to the heavens is circular: celestial bodies rotate around the centre of the universe (which is the centre of earth), and never acquire a new place, contrary to sublunar bodies. In fact, every segment of a circular movement is recurrent in time and equal to itself, there is no natural place for celestial bodies for they are not directed anywhere, and simply repeat the same movement. From a different perspective, finality is again not applicable to celestial bodies. In this paradigm, circular movement is thus the most perfect of movements, since it is endless and not directed. It is then a movement of a different nature from rectilinear movement (downwards and upwards), which is proper of sublunar bodies. With regards to circular movement, King 1624 writes that:

> Motus circularis non fit ad terminum in quem exeat, sed recurrit in sese, et partium tantum est totius quiescentis, quieti simillimus.
>
> Nec incipit nec desinit, sed in se reflexus recolligitur, continuitate sua uniformis; etsi durationis, et spatij terminis nullis definitur. [TP XII]

In the *Theses philosophicae* we never find a single argument taken as the principal argument for the demonstration of such difference in nature, contrary to what happens in the case of prime matter. In fact, from the theses written in 1629 in St Andrews, we know that regents favour the argument for the existence of prime matter which is based on natural philosophy alone, which is considered stronger than others precisely in virtue of its purely natural philosophical nature.[8] Rather, the demonstration of the difference in nature can be obtained from different perspectives, all of them equally valid as starting points, all of them equally valid as background theories, depending on the case. The empirical evidence for the difference in the nature of the heavens is circular movement, which is absent from our experience of sublunar bodies in movement. Yet, circular movement alone cannot prove any of the properties of celestial bodies, just as, in the view of some regents, the nature of heavy and light bodies cannot prove (that is, it cannot be a middle term in a demonstration) the downwards or upwards movement, and only leads to prove mobility.[9] Circular movement becomes the empirical support for a number of theories supposed only proximately by this evidence. This does not entail any illicit passage; it simply shows how some Scholastic theories are the result of a number of mutually sustaining premises.

[8] I analysed this argument in part I, chapter 1, section 2.1. The argument is labelled *'ex naturali rerum generatione'*.

[9] Strachan 1631, TP IV.1. Part II, chapter 2, section 3.3.1.

Recent philosophy of science has shown, thanks to the works of T. Kuhn, P. K. Feyerabend and I. Lakatos among others,[10] the strength of paradigms in shaping the philosophical world and in somehow validating or refuting evidences and theories. Moreover, the very notion of 'empirical evidence' and 'proof' seems to be weaker than commonly believed. Sibbald 1623 makes the only reference in graduation theses to a very recent innovation in astronomy, destined to dramatically change natural philosophy:

> Coelum recte statuitur quinta essentia, ab elementis distincta iisdem nobilior.
>
> Nec contrarium ex optica demonstrari potest. [TP 19-20]

It is clear that the regent is referring to the telescope. It is possible that Sibbald read of it directly from Galileo's *Sydereus Nuncius*, published in 1610. As a matter of fact, the regent refers to *"Iohannes Pena et alii"* as supporters of this view. It seems that Galileo's reasons did not convince Sibbald, who rejects the idea that optics can play a role in discovering the nature of the heavens, or better, in changing what we know of the nature of the heavens. We have thus evidence of an endorsement of traditional cosmology after the beginning of the so-called scientific revolution: Sibbald does not expand his point any further, but we can argue that he would favour the vast body of Scholastic literature supporting the quintessence doctrine over the observations of Galileo.

What Sibbald may not have favoured is a theological interpretation of the quintessence doctrine. Another unique passage is found in King 1624, who, in a way uncommon in the seventeenth century debate, exploits the biblical reference to Joshua fighting against the Amorites:

> And the sun stood still, and the moon stayed, until the people had avenged themselves upon their enemies. *Is* not this written in the book of Jasher? So the sun stood still in the midst of heaven, and hasted not to go down about a whole day.[11]
>
> Non solum Sacrae literae, quae testantur [symbol of sun] pugnante Iosua 3. horis constitisse, ad orationem Hezekiae 15. grad. regressum esse, Stellam novam Magis apparisse: sed etiam novorum syderum

[10] For example, I. Lakatos, *Philosophical Papers of Imre Lakatos*, 2 vols., J. Worrall - G. Currie (eds.), Cambridge, Cambridge University Press, 1978; P. K. Feyerabend, *Against Method*, London, Verso, 1993 3rd edition; T. S. Kuhn, *The Structure of Scientific Revolution*, Chicago and London, University of Chicago Press, 1996 3rd edition.

[11] *King James Bible*, Book of Joshua, 10:13.

> procreatio, unius, Anno 1600, in Cygno juxta eam stellam quae in ejus pectore lucet, alterius, quod Anno 1604, in [symbol of Sagittarius] visum est: Cometarum etiam in Aetherea regione supra [symbol of moon] situs, coeli mutabilitatem arguunt. [TP XIV]

King is quoting the passage in Joshua to prove that the heavens are mutable: not simply 'in movement', but mutable, which means that God can create new stars, or change the position of stars by means of his absolute power. The use of this passage is interesting for three reasons: 1) in the struggle between the Roman Church and some philosophers and scientists, such as Galileo, the sun stopping to allow the Jews to win their battle was usually mentioned on the side of geocentrism, as a proof that the sun is orbiting around the earth. King instead employs it as biblical proof of the mutability of the heavens. 2) The biblical passage is quoted alongside recent astronomical observations:[12] both the Bible and experience, according to King, convince us that heavens are not immutable. Yet, King is the same regent I quoted regarding circular movement: in his philosophy a quite innovative acceptance of the mutability of the heavens does not entail identity of nature between the heavens and the sublunar world. 3) The Bible is regarded as a source of information about the universe: this is, again, unique to this passage for its explicitness. Regents hold that the Bible provides support for philosophical doctrines when philosophy might be in conflict with revelation. A question arises as to what disciplines this conflict extends to, and natural philosophy is usually respected in its autonomy. Nonetheless, it is a fact that the heavens (as much as the relation between accident and substance) cause debates which call theology into question.

2. Movement of the heavens

Celestial bodies are of a different nature from sublunar bodies, with all that is thereby implied: no finality, no generation and corruption, no natural places, no four elements. Yet, celestial bodies do move, and regents dedicate much attention to the analysis of this movement. Celestial movement seems to be local movement: this is proved by the fact that local movement is the only movement which does not entail a change in the moving

[12] King refers to event of 1600 and 1604, two supernovae explosions (*stellam novam*), the latter also recorded by Kepler. For a survey of the cosmology of the theses: J. L. Russell, *Cosmological Teaching in the Seventeenth-century Scottish Universities, part 1*, in Journal for the History of Astronomy, V (1974), pp. 122-132.

substance, a change that is impossible for celestial substances. Change in local movement is still a categorial change (in the category of *ubi*) but it is somehow extrinsic to the moving substance, which can change presence in space or whereness without a change in its (other) accidents. Scholastics hold that by movement alone no new relation to things is acquired, as we are reminded by Fairley 1623, TL III, commenting on *Phys.* V, 2. A new relation is established when there is a change in a substance, since a relation is an accident in a substance: local movement does not bring about any change in a substance, so no new relation either.

Local movement is predicated of celestial bodies, and it is the only type of movement which they share with sublunar bodies. The nature of this movement raises questions about: 1) the applicability of the principle *omne quod movetur ab alio movetur*; and 2) the possibility of movement in a void.

2.1 The principle of movement of the heavens

The principle *omne quod movetur ab alio movetur* plays a central role in the analysis of the movement of heavens, as much as it did for the movement of heavy and light bodies. Everything which moves is moved by something else: this 'something else' does not necessarily have to be an external cause, as we have seen that animate bodies and (at least according to some regents) heavy and light bodies do move themselves. In those cases, form as nature is what moves the substance. For inanimate bodies, such as a stone, the mover is easily identified with the external substance setting the stone in motion. The question is about what model applies to celestial bodies.

Regents almost unanimously hold that celestial bodies do not move themselves: there is no inner active principle of movement, and in particular the form of celestial bodies is not the principle of such movement.[13] They are instead moved by an external cause, the 'intelligence' (*Intelligentia*, identified with angels), which acts on the inner passive principle of celestial bodies, so that their movement follows their nature and is not violent. The role of this intelligence will be fully appreciated later on, when dealing with the finality of celestial movement: in fact, the regents respect the Scholastic principle that finality is always connected with an intellect which apprehends the end as good. King writes about the intelligence in a passage of his theses in 1616:

> Unanim<i> Philosophorum consensu, Coeli motus fit ab intelligentia, quae est substantia immaterialis Coelo assistens, libera et voluntaria intellectione movens.
>
> Motus Coeli non est pure naturalis, sed potius voluntarius: nec data est Coelo forma naturalis ad movendum ut perficiatur, sed forma voluntariae intellectionis ad movendum [...] [TP VII]

The heavens only have a sort of inclination towards movement, so that an external mover is required for them to be in movement. Other regents call this movement 'above nature' (*praeternaturalis*), not in the sense that it is unnatural (that is, violent) but simply that it is of a different type from sublunar movement. King also claims that the heavens do not move in order to acquire a greater perfection (unlike sublunar bodies), rather, Rankine 1631 states that *"coelum moveri ut moveatur"* [TP XIV.4], an expression identical to one used by Lesley 1625 that fire *"movetur ut moveatur, non ut quiescat"* [TP XIV]. In both cases, movement is conceived as a natural state for the heavens and the element fire, for both the heavens and fire do not move towards rest.

Intelligence is the principle of the movement of celestial bodies both as cause of their movement and explanation of their movement, as Robertson 1596 claims:

> Coelum materia est in se actuata. Non differt itaque coelum a natura coeli. Natura coeli, medium demonstrationis motus coelestis de coelo esse nequit: sic enim non differet medium et subjectum. [TP 11]
>
> Medium demonstrationis motus coelestis est intelligentia. Medium demonstrationis motus coelestis est causa externa, quoad informationem: nisi quis putet assistentiam causam internam constituere. [TP 12][14]

A consequence of the different nature of the heavens is that matter is completely actualised: the heavens do not differ from the nature of the heavens, while sublunar bodies do differ from their nature, and this difference triggers movement towards a greater perfection. Heavens' matter is not in potency, it has no appetite towards form other than its current form. Robertson claims that this is the reason why the nature of the heavens cannot be the middle term of the demonstration of the movement of the heavens: if it were, the middle term and the subject would be one and the same. In other words, we would be

---

[13] See also, for example: Craig 1599, TP 6; Wemys 1612, TP 16.II; Forbes 1624, TP XI; Seton 1627, TP XXXII.

[14] A translation of the *Theses physicae* of Robertson 1596 is in the Appendix.

explaining the movement of the subject (heavens) by means of a middle term identical to the subject itself: this would hardly give any explanation. Thus, the middle is the intelligence, an external assisting cause. In sum, intelligence has a threefold role with regards to the heavens: 1) as a metaphysical cause; 2) as an epistemological principle; 3) as providing finality [section 3].

2.2 *Resistentia medii* and void

> *Vacuum* vero, quia rerum unionem, et naturas destruit, ipsa Natura maxime abhorret: nec si daretur, ullus esset in eo motus. [Forbes 1624, TP IX]

Natura vacuum abhorret is a famous principle of Aristotelian and Scholastic philosophy in general: it is not exclusive to these philosophies (Cartesianism for example) and it is not a necessary principle, since some Aristotelians and Scholastics (including some regents) did not exclude the possibility of a void. Yet, the vast majority of Aristotelians and Scholastics considered that a void would be a dangerous breach in the fabric of reality, for it breaks down physical continuity and contact between substances. Forbes's passage can be taken as representative of this position. Scholastics hold that there cannot be action at a distance, which means, an agent always acts either through a medium which somehow conveys the causal power of the agent, or through direct contact with the patient. The presence of a void (which is the absence of substance) would inevitably interrupt this chain of causality, making natural causality ineffective. Later on in the seventeenth century, one of the innovations of Newtonianism will be a picture of reality in which void as a place and action at a distance (i.e. gravity) are intelligible.

Many Scottish regents seem to accept the notion of a void and integrate it into their philosophy. Their talk about a void usually has three options: 1) a void is not natural and cannot exist; 2) a void is not natural, yet we can speculate on what would happen if it existed; 3) void is natural and it exists. Options 2 and 3 are most common in the theses, and Forbes 1624 can be said to have submitted a minority report.[15]

[15] For example, the theory that a void is unnatural and that it does not exist is held by Forbes 1624, TP IX and, perhaps, Rankine 1631, TP VIII. A variation of this theory, that a void is unnatural and that it does not exist, yet that we can speculate about a movement occurring in it, is held by Adamson 1604, TP 2; King 1612, TP 10; Fairley 1619, TP VI; Sibbald 1623, TP 12-13; Lundie 1627, De vacuo seu inani TP II; The third theory, that a void is natural and that it exists, is held by Reid 1614, TP 12; Lesley 1625, TP IIII; Stevenson 1625, TP XVI; Wemys 1631, TP IX.

Aristotle makes a direct connection between movement and void. In *Phys.* IV, 8 he claims that in a void the local movement of a substance would be infinitely fast, since no substance would resist the moving substance. With no opposition, the substance would move at infinite speed, since according to Aristotle the movement of a substance is the result of the impetus contrasted by the resistance of another substance. It appears that every movement is brought about at a finite speed, and we have no perceptual experience of a void: then, this infinite speed is impossible. It appears that the absence of a void is the condition for movement to occur as experience shows that it does, with finite speed in a finite period of time. This consideration includes both sublunar and celestial bodies, all identical when it comes to local movement in a medium. Thus, void is rejected on two grounds: 1) movement would occur at an infinite speed due to no resistance by the medium; 2) a void would bring about gaps in the natural world.

I think that these considerations are most intelligible when referred to celestial bodies. Contrary to sublunar bodies, celestial bodies do seem to move at a regular and constant speed in an empty environment (the heavens), evidence which pushes regents to open up to the idea of void and to rewrite the Aristotelian theory of resistance of the medium. These are Fairley 1619's words on the matter:

> Circularis Coeli motus est continuus et successivus, cum tamen fiat absque ulla resistentia ex parte medii.
>
> Resistentia ex parte medii, quae est extrinseca, non requiritur necessario ad motum localem. Alia igitur ratio est successionis in motu locali, eaque duplex, scilicet latitudo distantiae in medio repertae seu intercapedo et distantia extremorum, ac latitudo extensionis ipsorum corporum ob quam repugnat partes priores et posteriores simul praesentes esse eidem puncto aut parti spatii.
>
> In vacuo, si daretur, non modo fieret motus localis, sed et in tempore. [TP VI]

Fairley starts from the evidence of the regular movement of the heavens despite the absence of a medium. In the sublunar world, where we have no evidence of void, the regularity of movement can be referred to the regular resistance of bodies, so that a body can move in a medium according to its impulse (how strongly it is in movement) and to the resistance of the medium (how strongly it is contrasted). In the heavens there is no medium: thus the perceived regularity of celestial movement (indeed the most regular of all movements) must be accounted for according to some other principle.

This is Fairley's argument: resistance is not a necessary principle. Movements do occur in the absence of a medium. A body in movement (whether sublunar or celestial) is necessarily extended in space and the distance it covers while moving is necessarily an extension in space as well. It is impossible that parts of the moving body occupy the same portion of space; that means, a moving body necessarily retains its internal division and proportion between parts. Therefore, the sufficient principles of movement are the spatial extension of moving bodies and the spatial extension of the space in which the movement occurs: every time there is extension in space, then movement is successive and regular, and also in time, not instantaneous. Fairley calls this extension in space 'latitude' (*latitudo*). Resistance of the medium is an external principle which concurs with movement, but is not the condition for movement.

In other words, we can imagine a body in a space, moving from point A to point B. The moving body is itself extended in space, because no natural bodies can be without extension in space. The distance between A and B is a finite distance, just as the extension of the moving body is finite: no infinite bodies or distances can exist in nature, according to Aristotle. The body will move from A to B in time, with a regular and successive movement: in this picture, movement is about extension and dimensions of the moving body and of the distance covered, it is not about a proportion of resistance of the medium and impulse of the moving body. Even in an empty space, distances retain their value, and distances cannot be overcome except over a period of time.

I believe that Duns Scotus influenced those regents who accept this theory of movement. Scotus in fact surpassed the Aristotelian account of movement in a void by claiming that the sufficient condition for regular movement is distance, not plenum.[16] Lesley 1625 compares the Aristotelian and Scotistic versions, and then expounds his own theory:

> In natura vacuum non est, 4. *Phys.* in quo et, si esset, non esset motus; qui cum omnis fiat in tempore, *ibid. t. 129* adeoque tempore sit continuus, *ib. t. 99.* absque pleni resistentia nullus est, *Averr. 4. Phys. com. 71. et seqq.* Resistentia, in qua, medii externa: quippe interna, quam ponit *Scot. 2 Sent. dist. 2. q. 9.* nulla, nisi κατὰ συμβεβηκὸς, *Zab. I. de Mot. Grav. 12.* Atqui in natura vacuum est; et si non esset, non esset motus: cujus quasi principium est vacuum, quod cum semper sit plenum, fit vacuum, ut impleatur, *Scal. ex. 5. n. 2.* [TP IIII]

A very interesting passage. In Aristotelian philosophy, a void breaks down the relationship between time and movement, because without resistance of the medium

[16] On the Scotus's theory of void: A. Broadie, *Duns Scotus on Ubiety and the Fiery Furnace*, in British Journal for the History of Philosophy, 13 (2005), No. 1, pp. 3-20, in particular pp. 12-13.

movement takes place in an instant: and this is in contradiction with the principle that 'everything takes place in time'. Lesley mentions Scotus's theory of the internal resistance of the moving body, the very theory which Fairley accepts: this internal resistance is a principle only by accident, as Zabarella claims, because an external principle is required as well. Once again Scaliger is quoted with approval, and Lesley takes his own theory of movement from Scaliger's *Exercitationes*. The void is said to be a 'quasi-principle' of movement, because it makes movement possible by being filled by a body moving into an empty space. Were all space filled (that is, occupied by substances) no movement would occur. Lesley claims that void does not exist in nature as an empty dimensional space, but that it is immediately occupied by a substance. He retains the finalistic principle that nature rejects void, and that somehow void exists in order to be filled by substances. What can be said with regards to both theories is that a new *ubi* is a mode intrinsic to the moving body, independent of void and plenum.[17]

Lesley's theory is different from Fairley's. Both accept the notion of void: Fairley seems more familiar with the Scotistic idea of an empty dimensional space, while Lesley still holds that all reality is a plenum. This is why he rejects the Scotistic notion of internal resistance. These are, therefore, two different theories, which have in common the idea that void has a role to play in nature.

3. Finality of the heavens

The heavens are all the celestial spheres which surround and contain the sublunar world, and all the substances within: inevitably, talking of 'finality' of the heavens is talking of finality of the universe as a whole. We have seen that celestial bodies do not undergo movement in the way sublunar bodies do: there is no such directed change towards a new form intended as the end of change. Celestial bodies are what they have to be, the only change which affects them is the change in whereness (*ubi*). King 1620 even downplays this change in the celestial movement by saying that the proper terminus is not a new *ubi* but simply a new mode of whereness (*modus ubicationis*) of celestial matter: this way, the difference from sublunar bodies is even stronger, since no categorial talk is accepted.

[17] *"Terminus quem latio per se requirit non est locus, sed ubi, qui modus est quidam intrinsecus, et independens a pleno et vacuo."* King 1612, TP 10.I.

The question about finality of the heavens is similar to that about heavy and light bodies: how it is possible to account for the evidence of finality in respect of inanimate bodies.[18] Some regents endorse the Thomistic view that heavy and light bodies are directed by the mover (*generans*) who gives them such and such forms which determined whether they are heavy or light, as we are reminded by Stevenson 1625, in good Aristotelian fashion:

> Movens semper secum fert aliquam formam quae sit principium et causa motus. 3. Phys. 2. [TP XIV]

Other regents hold that finality is found within forms themselves of heavy and light bodies, asking why we cannot conceive of a model for heavy and light bodies' movement similar to that of animate bodies.

There is no such debate with regards to the heavens: regents unanimously claim that intelligence moves the heavens, so whatever finality the heavens show or act towards, is from the intelligence which moves them. The presence of intelligence as principle allows the regents to avoid the problem of finality because the model 'intelligence-finality of the universe' is structured on the basis of the model 'intellect-perceived good', proper to the analysis of human being. In fact, a perceived good always requires an intellect which perceived the good as such, and consequently moves towards it. For heavy and light bodies Strachan 1631 tried to introduce the notion of metaphorical intention, which is commonly used by Scholastics to express the sort of causality that a final cause has.

The heavens have no 'internal' finality, regents say:[19] they do not move towards any greater perfection than the one they already possess. If they did, they would not be different from perishable bodies. Yet, they move to the advantage of sublunar bodies: the endless vicissitude of generation and corruption is the 'external' end of the heavens' movement.

> Non movetur coelum totum, nec ulla ipsius pars propter sui conservationem; nihil sane acquirit novi propter se.
>
> Quare moveri propter nostram generationem putandum est. [...] [Reid 1610, TP 11]

Sibbald, regent at Marischal College in the 1620s, in two sets of theses, 1623 and 1625, puts forward his own interpretation of celestial movement, which involves the finality of

[18] 'Evidence of finality': none of the regents doubts that finality is apparent and omnipresent, and determines what the universe is like.

[19] For example, Sibbald 1625, TP IX, A quo coeli moveantur: *"Quod nimirum motu illo circulari nullam perfectionem intrinsecam, et debitam sibi adipiscantur, cum nulla tamen forma active inclinet ad motum, nisi per illum acquirenda fit aliqua mobili debita perfectio."*

movement. The regent argues against the possibility of proving that intelligence moves the heavens:

> In coelis nullum est vitae indicium, praeter motum localem qui seclusa cognitione et amore per se vitam non arguit.
> Coelum non est animatum.
> Coelum non ab intrinseca forma, sed ab extrinseco moveri demonstrare nulla ratio potest. Probabiliter tamen ostenditur ab extrinsecis motoribus cieri. Hi intelligentiae sunt. [1623, TP 24-28]
>
> At cum generans se moveat dum generat, non ob perfectionem suam, sed speciei debitam, omniaque ad omnes positionum differentias motu vacui moveantur, cur non coelum a seipso propter conservationem universi potest moveri? [1625, TP X]

In both sets of theses, Sibbald raises the doubt regarding the role of intelligences: he claims that it is simply 'more probable' that it is in fact an intelligence which moves the heavens. Sibbald is the same regent who rejects optics as a useful discipline in enquiring into the nature of the heavens: his overall theory of the heavens is not against Scholastic tradition. It is perhaps more interesting that a regent like Sibbald conceives of the hypothesis that the heavens move themselves, in an attempt to bridge the metaphysical gap between celestial and sublunar bodies.

4. Aristotle on the eternity of the world and the demonstration of the prime motor

Aristotle is without any doubt the main inspiration for the regents. The *Theses philosophicae* are often structured as commentaries on Aristotle's doctrines, he is ostentatiously quoted in Greek and his authority is required in almost all philosophical contexts. This is not surprising evidence: regents were teaching during a period of Scholastic renaissance and Aristotelian vigour (the two aspects do not always go together) following the Humanist reformation of philosophy. It is hard to say if Scholasticism prevails over Aristotelianism or vice versa in the theses: I believe that in this case the question is rather what interpretation of Aristotle the regents bring forward. I intend to address this point in this last section and then at the beginning of the Conclusions. I identify two approaches: in the Conclusions, I deal with the reception of Aristotle in general, seeking to show, in particular, in what cases the regents expound a Christianised version of Aristotle and whether we can conclude that they ultimately endorse an Aristotelian theory of substance. I will argue that Aristotle does not appear to be a cause of traditionalism in the Scottish universities: rather, as the case of the rejection of Transubstantiation shows, in the name of Aristotle regents went beyond contemporary Scholasticism.[20]

In this section I deal with a particularly interesting aspect of Aristotelian philosophy, which, I believe, is revealing of deep motives behind the philosophy of the regents: the interpretation of the principle *omne quod movetur ab alio movetur* and its role in the proof of the existence of the prime motor.

Most famously, Thomas Aquinas introduces his five ways for the demonstration of the existence of God by the principle that 'everything that moves is moved by something else' [*ST*, I, q. 2, aa. 1-3]. This also seems to be on Aristotle's mind in book VIII of the *Physics*, which leads to the proof of the necessity of a prime motor. Despite the fundamental difference between the two deities (Thomas's God is the giver of essence and existence, Aristotle's prime motor is the final and efficient cause of the movement of the world), the principle by which these two conclusions are reached is the same. Scholastics hold in fact that in respect of each of the four kinds of cause, material, formal, final and efficient, there is a first cause. Regents do not disagree with this fundamental point: we have seen that the existence of prime matter is also proved, a priori, by appealing to the existence of a first cause in the genus of material causality.

The validity of this principle rests on the assumption that an infinite regress is not a valid option:

> Ad primum ergo dicendum quod, cum omne quod movetur ab alio moveatur, quod non potest in infinitum procedere, necesse est dicere quod non omne movens movetur. [*ST*, I, q. 75, a. 1, ad 1]

In a series of efficient causes, the latest effect is caused by its immediate cause, which is itself the effect of its immediate cause, and so on, to infinity. The logical problem is that in order to have the latest effect we also must have an infinite series of causes, which ultimately make it possible for the latest effect to be actual. Yet, an infinite series cannot be

[20] A similar consideration is made by I. Düring (*The Impact of Aristotle's Scientific Ideas in the Middle Ages and at the Beginning of the Scientific Revolution,* in Archiv für Geschichte der Philosophie, 50 [1968], pp. 115-133) regarding the appreciation for the "real" Aristotle, rediscovered by modern philosophers and scientists (p. 129). Regarding the Scottish context, the opposite theory is found in R. S. Rait, *Andrew Melville and the Revolt against Aristotle in Scotland,* reprinted from The English Historical Review, London, April 1899.

actual in time, because it is always possible to posit an ulterior cause further back in time. This is why Thomas tackles the problem without the temporal succession.[21] His proof deals with a series of contemporary causes all concurring to the existence of the latest effect (which is not 'latest' in time): if we imagine a man throwing a stone, the series of present causes leads us up until God, first efficient cause. This is also why Scholastics hold that the difference between creation and conservation of the world is only a distinction of reason: in Descartes' narrative, God's activity is constantly required.

Now, in Aristotle's philosophy, the world is believed to be eternal: there is no concept of 'creation from nothing', the divine is the principle of "organisation" of an eternal world. The concept of creation made its entrance in philosophy during the first centuries of the Christian era, thanks to the thinking of Philo of Alexandria, Philoponus, the fathers of the church and the late Platonists. A profoundly influential change in the philosophical interpretation of the world. Thomas believed that, on purely philosophical ground, we must commit to Aristotle's conclusion that the world is eternal.[22] Our natural reason alone cannot decide against it, nor can it decide for it. Yet, creation in time is philosophically possible, and revelation tells us beyond any doubt that the world was created in time. After the acquisition of this truth by means of revelation, natural reason can find arguments in its favour and can show that revelation is not in contradiction with reason.

Where do regents stand in this grand debate, just briefly sketched here? Regents are Christian Reformed philosophers, they believe in the Christian revelation and this faith is reflected in their philosophy. As I had occasion to point out earlier, the natural philosophy of the *Theses philosophicae* is consistently regarded as an autonomous discipline, where the appeal to God's intervention is very limited. To be more precise: God is the ultimate and first warrant of the order and existence of the universe by its *potentia ordinata*, no regents would deny this; where they stand away from Catholic Scholastics and a number of modern philosophers is in their search for an explanation of the created world without involving God's *potentia absoluta* or a reiterated divine intervention in the natural course of events: the only example is the creation of the human soul at the moment of conception. We have seen that regents reject the miracle of Transubstantiation in the graduation theses not on the basis of biblical authority but on the basis of, as they say, 'good philosophy'. It is arguable, and I believe it is correct, that both a Protestant reading of the Bible and an understanding of Aristotle on the relation between a substance and its accidents make them inclined to find

[21] The example is that of a hand that moves a stick that moves a stone: three movements in a causal sequence, and perceptibly simultaneous.

[22] As in Seton 1627, TP XXX: *"Creatura secundum naturam suam potuit esse ab aeterno"*. The eternity of natural species is an Aristotelian theory, which Seton accepts as a conclusion of a purely philosophical enquiry.

philosophical arguments to deny the miracle of Transubstantiation: it is remarkable that they achieve such a rejection by ostensibly appealing to philosophical arguments.

Similarly, a significant number of regents explicitly reject both the truth of the principle *omne quod movetur ab alio movetur*, and also the conclusions based on this principle.

Regarding the eternity of the world, Reid 1622 writes that:

> Si nulla forma introducatur nisi ex materia privata, ex Arist. qui mundum falso aeternum esse putavit, utrum forma privationem, an privatio formam antecedat, nequit determinari.
>
> At ex veritate, qua nos Christiani mundum a DEO ex nihilo conditum fuisse credimus, absolute loquendo, forma tempore etiam praecessit omnem privationem Physicam et particularem. [TP VI]

Reid's idea, also stated in 1610, is that natural reason alone cannot prove whether form or privation came first in time; which means that Aristotle was wrong by his own logic in believing the world eternal. Christian revelation tells us that the world is created in time, which means that form precedes privation in the series of generation and corruption: first there are substances, then the beginning of the series of corruption and subsequent generation (*unius generatio est alterius corruptio*), a series which is posterior by nature to the creation of substances. When natural reason stops, revelation provides ground for finding truth.

Two regents are particularly clear in rejecting book VIII of the *Physics*: King 1612 and, again, Sibbald, in his theses of 1623.

> Omne agens ex naturae necessitate secundum ultimum suae potentiae gradum, ac tantum quantum potest, agit.
>
> Deus igitur, cum sit infinitae virtutis, nec effectum produxerit infinitum, non agit necessario.
>
> Quum itaque illa Aristotelis opinio de mundi aeternitate his duobus principijs innitatur tanquam fundamentis, necesse est ipsa etiam corruat, adeo ut mundus etiam a Deo in tempore creari potuerit, vel ex principijs Philosophiae. [King 1612, TP 16]

> Propositio haec, Omne quod movetur, ab alio movetur, aut falsa est, aut licet vera infirmum nimis fundamentum demonstrationis primi motoris.
>
> Verius et evidentius principium illud Metaphysicum: *Quicquid fit, ab alio fit.*

> Recte Avicenna non Physici, sed Metaphysici esse demonstrare ens dari aliquod primum et increatum. [Sibbald 1623, TP 14-16][23]

King and Sibbald attack Aristotle from two different viewpoints, and reach the same conclusion: the Aristotelian theory of the prime motor is ill-based.

King focuses on the powers of an agent. Even if the agent is of infinite power (*virtus*) like God, it does not act by necessity and does not produce an infinite effect. This is explained by the notion of the free act of creation and by the impossibility of an infinite (created) being, as Aristotle himself would confirm (King quotes *Met.* VIII on this matter, few lines above, TP XV.1). According to King, Aristotle's demonstration of the eternity of the world is precisely based on these two wrong assumptions; which inevitably make the conclusion wrong as well. Even according to Aristotle's principles then, creation in time is possible [TP 15].

Sibbald includes in his criticism of Aristotle the very principle *omne quod movetur ab alio movetur*, regarded as evident and solid by traditional Scholastics. He presents two possibilities: either 1) the principle is false; or 2) even if it is true, it does not provide ground solid enough for Aristotle's demonstration of the prime motor. Both possibilities imply a rejection of the relevant passage in book VIII of *Physics*. I believe that an antecedent of this position can be found long before Scholastic philosophy, during the very initial moments of the appropriation of the Christian revelation by philosophers: in the *De Aeternitate Mundi Contra Aristotelem* by John Philoponus.[24]

Philoponus's original books have long since vanished. His ideas on the eternity of the world are now known to us because of the polemic he started with Simplicius, who transcribed long passages by Philoponus in his reply to him.[25] Philoponus opposes Aristotle on many physical doctrines. What matters here is book VI of his *Contra Aristotelem*, where he sets out to criticise the arguments for the eternity of movement and where he puts

---

[23] To be contrasted with Wemys 1612, TP 13.I: *Primi ergo motoris in 8. Phys. ex motu primo demonstratio Physica est, non Metaphysica."*

[24] Philoponus, *Against Aristotle, on the Eternity of the world*, edited by R. Sorabji, translated by C. Wildberg, London, Duckworth, 1987; R. Sorabji, *Philoponus and the rejection of Aristotelian science*, London, University of London, 2010. On the Renaissance reception of Philoponus: C. B. Schmitt, *Philoponus' Commentary on Aristotle's* Physics *in the Sixteenth century*, in C. B. Schmitt, *Reappraisals in Renaissance Thought*, chapter VIII. The author underlines the fact that the commentaries by Simplicius and Philoponus provide a criticism of Aristotle which is not far from that offered by the 'new science' of the seventeenth century. Schmitt believes that the Renaissance re-discovery of Simplicius and Philoponus provided more arguments to the anti-Aristotelian philosophy and science. I believe that, at least in part, this is the case for the graduation theses as well, whose natural theology does not seem to be according to the Aristotelian principles.

[25] *Against Aristotle*, p. 24 ff. The two main works from which we can attempt to reconstruct Philoponus's theory are Simplicius's commentary on *de Caelo* and on the *Physics*.

forward the idea of creation from nothing. What is particularly interesting in Philoponus is that he criticised Aristotle from an early Christian viewpoint; this allows us to appreciate a reading of Aristotle before his "Christianisation" operated in the Middle Ages. Furthermore, Philoponus was never completely forgotten in the western Christian world, even if we must wait until the late sixteenth and early seventeenth century to see signs of growing interest in his philosophy.[26] Library records in Aberdeen university dating back to 1624, catalogued as MS M 70, show that at least one copy of two commentaries by Simplicius were available: precisely *Simplicius in quatuor Libros Aristotelis de Coelo*, published in 1527 by Aldus Manutius in Venice and *Simplicius in tres Libros Aristotelis de Anima*, 1527, for which no publishing place is noted. It is then probable that some of the Aberdeen doctors, for example Sibbald, would be acquainted with Simplicius's reports on and criticism of Philoponus.

Leaving the important archival evidence aside, I believe that Sibbald's short argument can be explained by Philoponus' criticism of Aristotle. In book VI of his *Contra Aristotelem* Philoponus argues against *Physics* VIII, 1 where Aristotle claims that if two bodies have not always been in movement, then there must be a movement prior to them, in virtue of which later movements occur. This is also true of this 'prior movement', so that it is impossible to posit a 'first' movement in time. Philoponus sets out to resolve this difficulty by the means of creation from nothing, which breaks the series of mover-moved bodies to reach a first absolute unmoved mover.[27] Philoponus's critical argument rests on the sequence of movements being in time: a qualification he ascribes to Aristotle and which Thomas, for instance, refutes in his own interpretation of Aristotle.

This might be what Sibbald has in mind when claiming that the principle *omne quod movetur ab alio movetur* is either false or insufficient to prove the existence of the prime motor. The principle appears to be valid only in natural philosophy, and Sibbald holds that proving the existence of the prime motor is a task of metaphysics, not of natural philosophy. Sibbald seems more sympathetic towards book XII of *Met.*, 6, 1071 b 2 - 1072 a 18, where Aristotle reaches the conclusion of book VIII of *Physics* in terms of act and potency (*"Verius et evidentius principium illud Metaphysicum:* Quicquid fit, ab alio fit.*"* TP 15). In natural philosophy it might be true that everything that is moved is moved by something

[26] As Sorabji interestingly points out, Philoponus is quoted by Galileo more often than Plato. *Ivi*, p. 2.

[27] *Ivi*, p. 131, fragments 117-120. The innovation of the arguments of Philoponus lies in pointing out an apparent flaw in traditional Aristotelianism and, in general, in the worldview of antiquity. Simplicius, among many others, argues for the eternity of the world; that means, an infinite number of years has passed until now. Philoponus points out the decisive contradictions: a world infinite in time contradicts the Aristotelian principle that nothing infinite can be actual, and an infinite number of years passed until now has to be increased, as more years follow from now on. R. Sorabji, *Philoponus*, pp. 213-214.

else, but it is also true that by force of this principle alone the natural philosopher cannot demonstrate the existence of a prime motor, and must instead limit their enquiry to the physical world.

How to interpret the open rejection of this principle in the light of the *Theses philosophicae* as a whole? No other regent is as clear as Sibbald on this subject; yet, contextual evidence can be given for what I believe is the very limited role for natural theology in the philosophy of the regents.[28] No proof for the existence of God is present in the theses until the 1650s: this includes proofs from our knowledge of the physical world. If we look further in the seventeenth century, we find an increasing interest in the Cartesian arguments of the *Meditationes de Prima Philosophia.* With the arrival of Cartesianism in the Scottish universities, the demonstration of the existence of God is, in 'Scottish Cartesian' fashion, a preliminary step to philosophical enquiry, alongside the argument of the 'cogito ergo sum'. This profound shift in exposition is striking. It is clear that Cartesian philosophy stimulated an interest for this argument which is missing in earlier theses: regents in the 1660s-1670s fully endorsed Cartesianism. The demonstration of the existence of God cannot be said to be the centre of heated debate in late Scholasticism, nonetheless it is a central part of most Scholastic works. The *Theses philosophicae* belong to the textbook Scholastic tradition, works written with the specific idea of providing an accurate yet not fully exhaustive account of philosophy, for the purpose of educating young students. The absence of this argument alone cannot lead us to definitive conclusions about the role of natural theology in the regents' natural philosophy. Nonetheless, this absence becomes more meaningful if interpreted in the light of an almost total absence of the discourse about God in natural philosophy. God's intervention is also denied (with the interesting exception of Dalrymple 1646) in the causality of secondary causes.

It might be the case that Sibbald makes explicit what is implicit in all other regents: the existence of God is not a subject of philosophy, it is a subject of theology and faith. John Calvin famously expressed his theory of the 'sense of god' (*sensus divinitatis*) according to which awareness of divinity and belief in God are well-nigh universal.[29] In Calvin a

---

[28] For the analysis of natural theology in Reformed Scholasticism: R. A. Muller, *Post-Reformation Reformed Dogmatics*, ch. 5. Muller indentifies natural theology as part of revealed theology, and claims that the idea that there is no role for it in Reformed Scholasticism is the product of later theology (such as Karl Barth's), and is thus foreign to the Reformers. In relation to the graduation theses, I think that one aspect is important: the distinction between philosophy and theology. If it is true that natural theology is part of revealed theology, it is also true that the development of natural theology was perceived as an excess in the direction of rationalism (*ivi*, p. 170). The graduation theses seem to belong to this faction of Reformed Scholasticism: the distinction between philosophy and theology is strong, and natural theology does not belong to the area of enquiry of the philosopher: in particular, of the natural philosopher.

[29] T. F. Torrance, *The Hermeneutics of John Calvin*, Edinburgh, Scottish Academic Press, 1988, in particular pp. 84 ff.

'Scholastic' demonstration of the existence of God is missing, and he prefers the Pauline doctrine that God is revealed in nature:

> For the wrath of God is revealed from heaven against all ungodliness and unrighteousness of men, who hold the truth in unrighteousness;
> Because that which may be known of God is manifest in them; for God hath shewed *it* unto them.
> For the invisible things of him from the creation of the world are clearly seen, being understood by the things that are made, *even* his eternal power and Godhead; so that they are without excuse. [*King James Bible*, Romans 1: 18-21]

It is then possible that regents reflect an approach to philosophy influenced by the Calvinist origin of their confession, in which the existence of God cannot be the conclusion of a philosophical argument. Mutatis mutandis, I believe that this position is consistent with the rejection of Transubstantiation: regents deny that philosophy can account for theological matters, either the miracle of the conversion of bread and wine into body and blood of Christ, or the existence of God. A matter of faith is not a matter of philosophy, even if faith always leads our philosophical interpretation of the world.

The *Theses philosophicae* until the 1650s do not commit to any discourse on God which is not either moral or metaphysical: God is present in philosophy, but natural philosophy is treated as a discipline independent of our knowledge of God, other than the faith in a benevolent, rational and free act of creation. Inevitably, the faith of the regents shapes their natural philosophy: they diverge from Catholic Scholastics with regard to the accounts of substance, extended matter, inherence of accidents and also the limits of natural philosophy: within a Scholastic philosophy, regents show a clear respect for the autonomy of natural philosophy. I believe that this is a clear example of the way in which the religion of the regents both influenced their philosophy, and also prepared the ground for the success of the scientific revolution in the Scottish universities in the later seventeenth and early eighteenth centuries.

5. Conclusion

The regents are still committed to the distinction in nature between the sublunar and the celestial world. Thereby, they reveal how deeply they are influenced by the tradition of Scholastic natural philosophy.

The celestial world is different in nature from the sublunar world because it is not subject to corruption. Sublunar bodies come to be and cease to be, while celestial bodies are eternal. Ergo, they are in movement, but in a different type of movement. Celestial bodies, for example, do not move towards an end in the way sublunar bodies do; more precisely, celestial bodies only have an 'external' end, which is the preservation of the sublunar world. If we abstract from this external end, the celestial bodies have no finality, that means, they are fully actual, and perfectly realize their nature.

I have called this theory of the difference in nature a 'paradigm', since the regents do not seek to prove it, but rather consider it as a starting point of their cosmology.

The heavens are moved by the intelligence; unlike the heavy and light bodies, the principle of movement of the celestial bodies is external. Yet, it is natural, because an internal propension towards a movement triggered by an external agent suffices to qualify such movement as natural. Unlike heavy and light bodies, celestial bodies are not the primary cause of their movement.

The regents make an interesting case for the movement of the celestial bodies in a void: probably influenced by Duns Scotus and against Aristotle, they claim that a movement in a void is possible, and takes place in time, because a moving body is extended in place, even in a void. They seem to accept the Scotistic notion of a void as a geometrical space potential occupied by bodies.

The investigation of movement has raised the question of the interpretation of the principle *omne quod movetur ab alio movetur*, traditionally exploited in natural theology. The regents seem to reject the validity of this principle beyond the physical world, and to rule out natural theology from their natural philosophy. I have argued that they might be influenced by a Calvinist form of Protestant philosophy.[30] In turn, this analysis has prompted the question of the reception of Aristotle, to which I turn now.

---

[30] As I have sought to prove, the philosophy of the regents is shaped by a form of Calvinism: in the cases of the definition of the accident and of natural theology, philosophical doctrines are rejected or approved on the basis of the Scottish Calvinism of the regents. Another interesting example of how Calvinism directly influenced philosophy is presented by C. H. Lohr in *The Calvinist Theory of Science in the Renaissance*, in G. Piaia (ed.), *La presenza dell'aristotelismo padovano nella filosofia della prima modernità*, pp. 123-132. According to Lohr, Calvinist philosophers distinguished themselves from both Catholics and Lutherans in terms of the conception of scientific knowledge, of the distinction of the philosophical disciplines and of the role of natural theology. Lohr ascribes to these differences the very origin of the idea of a "system" in Christian teaching, of an "organic" conception of knowledge and, ultimately, of the end of metaphysics as the "queen" of the sciences (p. 131). See also: C. H. Lohr, *Latin Aristotelianism and the seventeenth-century Calvinist theory of scientific method*, in D. A. Di Liscia - E. Kessler - C. Methuen (eds.), *Method and order in Renaissance philosophy of nature*, Aldershot, Ashgate, 1997, pp. 369-380.

Conclusions

1. Outline of the conclusions

The analysis of the cosmology of the regents has shed some light on the distinction between natural philosophy and natural theology in the theses. The regents seem to reject natural theology and the application of its principle 'omne quod movetur ab alio movetur' beyond the limits of the natural world. Most famously, Aristotle concluded the *Physics* with the application of this principle to the discovery of a first mover, regarded as the final and efficient cause of the universe. We have seen how regents put forward an interpretation of the principle which seems to exclude its use in natural theology, and, without the support of the Christian revelation, leads to the Aristotelian doctrine of the eternity of the world. Thus, it is in virtue of the revelation that regents go beyond Aristotle and hold that the world has a beginning in time.

The question of the reception of Aristotle in the theses is historically central: the importance of Aristotle is obviously not limited to natural theology. I shall here highlight two main aspects of the question:

1) Aristotle is the fundamental philosophical source of the theses. The result is the appropriation of Aristotle in a Christian philosophy. The two most debated doctrines are the eternity of the world and the immortality of the soul. I shall seek to enrich the previous discussion of these doctrines by reference to how the regents read the relevant Aristotelian texts.

2) We have seen in part I, chapter 4, section 2.2 that, according to Stevenson 1629, Aristotle and Porphyry do not accept the notion of an accident existing without a subject. In the first half of the seventeenth century, the passage in Stevenson 1629 is the only explicit connection between this theory and Aristotle that we find in the theses. We will see two more references later on in the century, by Forbes 1684 and Skene 1688, which will help to clarify how later regents looked back at the philosophy of their colleagues. It seems that, at least until the 1680s, the interpretation of Aristotle on this matter did not change, and that regents invoked the authority of Aristotle in the debate on the separate existence

of the accidents, prompted by the Catholic reading of the Eucharist. It follows then that also this later interpretation of Aristotle is in agreement with the Reformed Scholasticism of the theses.

In the final part of the Conclusions I shall expound the main aspects of each chapter, offering a general account of which can be considered the key features of the Scottish Scholasticism of the graduation theses. I shall finally seek to contribute to the answer to the question of the relevance of Scottish Scholasticism in contemporary research.

2. The reception of Aristotle in the *Theses philosophicae*

The philosophy of Aristotle is a major source of inspiration for the regents. The analysis of the reception of Aristotle is a preliminary question before drawing conclusions regarding the Reformed Scholastic character of the *Theses philosophicae*. As noted in part II, chapter 3, section 4, some regents are critical of the Scholastic principle that everything that moves is moved by something else, a principle which is traditionally used as a basis for the demonstration of the existence of God. This principle is also fundamental with respect to the Scholastic theory of movement, which entails that the natural state of bodies is rest. Bodies in movement naturally seek rest and tend towards it: in Scholastic natural philosophy, movement, not rest, requires an explanation. Therefore for every movement there must be a cause. As we have seen, regents do not reject this principle tout court: they reject a certain use of it. They believe that by the powers of this principle alone we cannot offer a demonstration of a non-empirical proposition such as 'God (or what we usually call 'god', to accept Thomas Aquinas's formulation) exists'. I argue that in rejecting that use regents commit themselves to a theory which is close to the doctrine brought forward by Philoponus against the Aristotelian Simplicius; namely, that the principle is to be understood in a temporal series of causes and effects, which can extend in infinitum, thus failing to provide the first cause in the natural series (which Thomas Aquinas, for example, believed himself to have provided).

A consequence of this theory appears to be the rejection of natural theology, understood as the attempt to prove the existence of God by means of our experience of the natural world: the regents' position fits well with the form of Calvinist confession they adhered to. In the Reformed Scholasticism of the theses, the proof of the existence of God based on the 'omne quod movetur ab alio movetur' principle is not subject to philosophical scrutiny.

This prompts the more general question of the reception and interpretation of Aristotle in the theses. The example of natural theology is perhaps the most evident sign of the fact that the reception of Aristotle is always followed by an interpretation of Aristotle, and consequently that the philosophy of the regents cannot be labelled "Aristotelian" without qualification.[1] In fact, even if we set aside the question whether Aristotle himself applied the 'omne quod movetur ab alio movetur' principle or not, it is a fact that within Scholasticism (just as within the Aristotelianism of the early Christian era) philosophers employed this principle in different ways, yet always believing their reading to be faithful to Aristotle.

The philosophy of the theses, if we accept this broad notion of Aristotelianism, is indeed Aristotelian. Aristotle is by far the most quoted authority; some theses are structured as commentaries on Aristotle's works; and he is always referred to with the utmost respect as 'the Philosopher'. One might object that these elements were features of philosophical writing and academic teaching widely standardised in Europe in the seventeenth century. Two considerations help to clarify the point: first, the acceptance of Aristotelian philosophy did not end with the arrival of the Reformation and the Renaissance reformation of philosophy. There is evidence of an enduring and successful Aristotelianism in post-Reformation Scotland, at least in the practice of university teaching. Along with the intrinsic philosophical merits of Aristotelianism, in the regents' eyes a Scholastic Aristotelianism was still the best pedagogical option, to such an extent that some scholars have suggested calling Scholastic philosophy in the late sixteenth and early seventeenth centuries 'academic Scholasticism'. Despite acknowledging the advantages of this formula, I believe that it overlooks the importance of Scholastic philosophy outside the academies: if it is undeniable that the backbone of Scholastic philosophy in the seventeenth century was an established academic practice, it cannot be forgotten that some of the greatest Scholastic works of the period were not directed towards academic teaching and exerted much influence in the public philosophical debate.

Secondly, I argue in section 2.1 that the regents' allegiance to Aristotle is also qualified, and regents were not afraid to interpret Aristotle in the light of what they believed was

[1] What I seek to provide here is qualification. From the 1970s on, thanks to the work of Charles Schmitt, Brian Copenhaver and others, scholars became familiar with the idea that each historical period had its own form of Aristotelianism. Schmitt thus suggested the expression 'Aristotelianism*s*' (*Aristotle and the Renaissance*, ch. 1) in order to account for the variety within Aristotelianism. I believe that E. Gilson held a different view on this matter: according to him, Thomas Aquinas, and more generally Thomism as a faithful interpretation of Thomas, is the true Catholic philosophy, and the best expression of Scholastic philosophy. This entails that the Thomistic Aristotle is the best possible interpretation of Aristotle for a Catholic scholar. I believe that Gilson knew the variety of the interpretations of Aristotle, along with the variety within Scholasticism, but only took one seriously: É. Gilson, *Descartes et la Métaphysique*

'good philosophy'. In section 2.2 I seek to outline the position of the regents in terms of their interpretation of Aristotle's theory of substance, which follows the debate on Transubstantiation: regents thought themselves good interpreters of Aristotle in claiming that it is impossible for an accident to exist without its natural substance. In sum, references to Aristotle are never just motivated by tradition, since Aristotle was still regarded as a powerful source of philosophical debate and progress.

A related question is about the relationship in the theses between Aristotelianism and Scholasticism. This question can perhaps be raised for any form of Scholasticism. In the theses, it appears that Aristotelianism and Scholasticism are much intertwined, to such an extent that it is impossible to detach one aspect from the other. Regents were Scholastic in the same terms as they were Aristotelian, and vice versa. Whatever interpretation of Aristotle the regents have, it is a Scholastic interpretation: whatever Scholasticism they have, it is an Aristotelian form of Scholasticism. We should not be misled by the Renaissance claims for the return to the 'authentic' Aristotle (which is part of the overall Renaissance attempt to return to the 'authentic' Classics), because in the seventeenth century in Scotland this claim was present in university teaching, but did not bring about a rejection of the Scholastic way in philosophy.

2.1 *Aristoteles Christianus*: Christian interpretation of Aristotle in the *Theses philosophicae*

As one might expect, the *Theses philosophicae* are not a case of an Aristotelianism which opposes Christian faith. All interpretations of Aristotle are kept within the boundaries of the Reformed religion of the regents: regents believe that the highest 'tribunal' for their philosophy is true religion, and they show no sign of the so-called 'doctrine of the double truth', as it is traditionally ascribed to Siger of Brabant. Yet, natural philosophy is indeed regarded as an autonomous discipline, but in no way can natural philosophy propose a truth which is incompatible with the Christian faith. When such conflict is evident, the regents resolve it in favour of the contents of revelation, either in terms of natural philosophical theories (as in the case of the rejection of Transubstantiation), or in terms of a 'suspension of judgement': a proposition which is left undecided in philosophy finds its answer in revelation (as in the case of the theory of the creation of the world in time). I believe that regents are truly Scholastic in this regard. The autonomy of natural philosophy

scolastique, Revue de l'Université de Bruxelles, No. 2, 1924 and *Introduzione alla filosofia cristiana*, is also coherent with this approach, insofar as natural philosophy is not understood as a 'mathesis universalis' which extends to the whole of philosophy: regents follow the Aristotelian principle that each philosophical discipline ought to follow the rules dictated by its subject-matter, and be defined by the limits of its subject-matter.

I also believe that the regents did not understand themselves as belonging to any philosophical school, for example, the Thomistic or the Scotistic school. What is true of most other countries in Europe in the sixteenth and seventeenth centuries, namely that philosophical *studia* and universities were structured either *in via Thomae* or *in via Scoti*, is not true of Scotland. I believe that the regents considered themselves as working within the Scholastic tradition, yet not bound to a specific form of Scholasticism. This said, the influence exerted by Duns Scotus cannot pass unnoticed: some major natural philosophical themes, such as prime matter as metaphysical act, formal/modal distinction, void as quantifiable extension, bear the mark of Scotus's philosophy. 'Eclectic Scotistic Reformed Scholasticism' seems an adequate description of the philosophy of the theses.[2]

In my analysis of the 'Christian Aristotle' in the natural philosophy of the theses I am not concerned with every theory in which the influence of Aristotle is felt. Most of the philosophy originates from the texts of Aristotle, and benefits from the long activity of interpretation and comment carried out from Simplicius and Philoponus onwards. What I have done is to offer an account of the most relevant passages in which the regents explicitly expressed reservations about the coherence of Aristotle with Christian revelation, and where the regents followed the practice of interpreting Aristotle as an ante litteram Christian philosopher.[3] This analysis can shed light on the question of which Aristotelian theories were understood as most in conflict with revelation, and which Christian doctrines Aristotle could hold on the basis of natural reason alone, Aristotle being the highest example of a philosopher unassisted by revelation.

From the viewpoint of a Christian natural philosophy, the two most debated Aristotelian texts are those regarding the eternity of the world and the immortality of the soul. With regard to the former, Aristotle held that the world is eternal, and that the first motor is the first final cause, not the Christian first efficient cause on which the whole existence of the

---

Milano, Massimo, 1986, forword.

[2] I believe that the qualification 'Scotistic' is necessary, and not included in a general account of Scottish Scholasticism as 'eclectic'. The graduation theses are a form of 'eclectic Scotism', rather than just, in general, a form of 'eclectic Scholasticism'. In fact, Scotism appears to be the thread linking all the graduation theses together, even if the regents' approval of Scotism is never uncritical.

[3] John Mair, in the liminary letter of his commentary to the Nicomachean Ethics, *Ethica Aristotelis peripateticorum principis*, Paris 1530, holds this opinion of Aristotle: *"Denique in tanto et tam multiiugo opere vix placitum unum Christiano homine indignum, si ut a nobis explanatum est legatum, offendas."* I owe this reference to Alexander Broadie.

world depends. Christian commentators of Aristotle perceived the problem of the absence of a theory of creation in Aristotle; this absence was usually explained by the claim that Aristotle's philosophy, as a purely human enterprise, had to stop where revelation was needed to provide further truth and advancement beyond philosophy. Thomas Aquinas, following Avicenna, developed a whole new metaphysics of the act of being which paved the way for a more mature interpretation of Aristotle within a Christian framework. An alternative solution is that the theory of creation was implicit in Aristotle's philosophy, even if Aristotle did not openly state it. This is what Forbes 1624 seeks to prove. After dealing with Aristotle's theory of the elements, the regent writes, regarding the universe, that:

> Quod Aeternum statuat, id licet homine Christiano indignum, Philosopho tamen Natura duce concedendum: quamvis verisimile sit, Creationem, qua ex Aeternitate, ut ipse putabat, universum condidit DEUS Aristotelem non latuisse: cum 12. Metaph. agnoscat Coelum et Naturam, a DEO pendere. [TP X]

The regent refers to book XII of the *Metaphysics*, perhaps to 6, 1071 b - 1072 a 20, where Aristotle proves the existence of an immaterial, eternal and immobile substance, mover of the universe. Forbes uses the Latin term *'pendere'*, which is philosophically ambiguous, since it may signify various forms of 'dependence', not simply the relation that a created universe has with respect to the creator. It is interesting that Forbes would read this passage, and arguably misread it, as implying a creative act by the first mover. His main argument is to be found in the preceding lines. Forbes holds that the eternity of the world is not an acceptable doctrine for a Christian man, and that some credit must be given to Aristotle, since he was solely guided by human reason. Yet, it is likely that Aristotle himself was not unaware of the possibility of creation, which is implicit in the description of the first motor provided in book XII of the *Metaphysics*.

A similar point is made by King 1612:

> Quumque inter solum nihil et aliquid, seu ens et non ens, sit infinita distantia: sequetur, vel ex principijs Aristotelis Deum ex nihilo aliquid creare potuisse. [TP 15.II]

This passage is part of a longer thesis which deals with the relation of an agent of infinite power (*virtus*) to its finite effect. An infinitely powerful agent can create either an infinite effect, or a finite effect in an infinite way (*infinito modo*). King claims that there is universal philosophical consensus that no creature can be infinite in act, as Aristotle himself claimed. The only option is therefore the creation of a finite creature in an infinite way. This is the introduction to the passage quoted above. In this passage King exploits the Aristotelian doctrine that being and non-being are opposed as contradictories, and between them there is an 'infinite distance'. This distance, namely, the possibility of a passage from non-being to being, is infinite because the opposition of contradiction between two elements is the strongest possible opposition, and it can be overcome only by an infinite power. In this context, 'non-being' and 'being' must be understood in an absolute sense: in the natural world, we only experience relative non-being and relative being. Natural generation and corruption occur between forms inhering in and informing prime matter, so that all forms can be said to be 'contraries' to one another in relation to prime matter, understood as the underlying principle of inherence, in the same way as colours are 'contraries' to one another in relation to the substance they are accidents of. In other words, prime matter is potentially open to forms, which are taken on successively by prime matter. This does not mean that two substantial forms can inform the same portion of prime matter: this is contradictory. It is not contradictory that two substantial forms inform the same portion of matter in temporal succession.

An infinite distance can only be covered by an infinite agent. There is an infinite distance between non-being (absolute nothing) and created being: the regent concludes that by the logic of Aristotle creation is possible. The following passage in King 1612 deals with the traditional principle *ex nihilo nihil fit*:

> Commune igitur illud Philosophorum classicum, Ex nihilo nihil fit, nedum ex principijs veritatis christianae, verum et ipsius Philosophiae, evertitur et corruit. [TP 15.III][4]

The regent does not ascribe this theory to Aristotle. Yet, it appears that the conclusion reached in TP 15.II by the logic of Aristotle implies the interpretation of the principle *ex nihilo nihil fit* only within the limits of natural philosophy, thus excluding the creation, which is a passage from absolute non-being to being.

The passages in Forbes 1624 and King 1612 are two different forms of the same attempt to credit Aristotle with, at least, the intuition of the philosophical theory of creation before

[4] Thomas Aquinas, *ST*, I, q. 45, a. 2, arg. 1: *"Videtur quod Deus non possit aliquid creare. Quia secundum philosophum, I Physic. antiqui philosophi acceperunt ut communem conceptionem animi, ex nihilo nihil fieri."* Thomas Aquinas is expounding here the theory of the Greek natural philosophers.

its historical formulation within the development of a Christian reading of Aristotle and Plato.

The theory of the immortality of the soul, which regents discuss with a direct reference to Aristotle, originates from their reading of *De Anima* III. Aristotle investigates the nature and the faculties of the soul, and concludes that a particular activity of the soul, namely, that of the agent intellect, which works out the universal on the basis of the impression on the possible intellect, is evidence for the immateriality of the soul. The external objects of our knowledge are individuals, and are perceived as such: Aristotle argues that the universal is the product of the agent intellect because the universal is not to be found in sensation. The work of the agent intellect is required in order to 'ascend' to the universal. Thus if the universal is not to be found in nature, then it does not come to our knowledge from nature. If our agent intellect can ascend to the universal, the agent intellect cannot be material; therefore, it is immaterial. The scholarly debate over the interpretation of this theory is vast: Scholastics favoured the interpretation that Aristotle either laid out the basis for the proof of the immateriality of the soul, or that he effectively proved that the soul is immaterial.

As noted in part I, chapter 2, section 1.1, the regents unanimously claim that the soul is immaterial, ergo immortal, since generation and corruption only affect material substances. The immortality of the soul is part of the Christian tradition, and it is no surprise that regents believe in it. What is more interesting is the argument deployed by, for example, Aedie 1616 in order to prove that Aristotle himself believed in the immortality of the soul.

> Philosophus. I. de Anima, cap. I. et 4. Tum etiam cap. 5. lib. 3. Animam dicit esse χωρίστην επί του σώματος, et I. cap. lib. 2. de Anima vocat μόρφην, et qua talem eam ibi definit.
>
> Immortalitatem igitur animae cognovisse et approbasse Philosophum constat.
>
> [...] Resurrectionem igitur mortuorum Philosophicis, quodammodo rationibus probabilem esse dicimus. [TP VII][5]

Aedie's reading of Aristotle is that the soul is 'separable' (χωρίστη) from the body and form (μορφή) of the body: therefore, soul is a separable substantial form. This means that soul is a form of the body but not a material form of the body; therefore the existence of the soul is not dependent on the existence of the bodily compound. The conclusion is that

[5] A translation of the *Theses physicae* of Aedie 1616 is in the Appendix.

Aristotle acknowledged the immortality of the soul and approved of it. The following step is the entirely non-Aristotelian notion of the resurrection of the bodies, a step dependent on the argument for the immortality of the soul. We have here a Christian reading of Aristotle. Setting aside the question of the faithful interpretation of Aristotle just as in the case of the eternity of the world, there is evidence that regents considered the Aristotelian passages quoted above by Aedie 1616 as convincing proof that Aristotle endorsed the theory of the immortality of the soul, and that he laid out the fundamental philosophical groundwork for such demonstration.

Another interesting example of a Christian interpretation of Aristotle concerns the doctrine of the unicity of the human soul. We read in Fairley 1615 that:

> Pluralitas animarum (ut de Theologia taceamus) in eodem composito, vel ex Philosophia Aristotelis absurda judicamus. [TP XXIV.1]

The heading of this thesis is the claim by Aristotle, in *De Anima* II, 3, 414 b 29-30, that the antecedent term is always included in the posterior, just as the vegetative soul is included in the sensitive soul, so that we must investigate case by case which is the soul proper to each species: a plant, a beast and a man (Aristotle's own examples). What the regent seeks to prove with this quote is that what is posterior (and arguably more eminent) includes what is anterior (and arguably less eminent), just as the rational soul includes both the vegetative and the sensitive souls.

All regents agree on the doctrine of the unicity of the human soul. We have seen in part I, chapter 3, section 1 that some regents hold the theory of the plurality of forms within the same compound. For example, regents take up Scotus's remark that the corruption of a human compound does imply two corruptions that occur in time: first, the corruption of the soul-body substantial union (what we properly call the 'death' of a man); secondly, the corruption of the bodily form-matter compound, that is, the dissolution of the body. The latter corruption is a process distinct from the former: the identity of the body with itself (the body of a dead man is still recognisable as the body of *that* dead man) does not depend on the union of the soul with the body, rather, on the union of the bodily form with matter. Therefore, two distinct substantial forms are present in the same compound. I argued that this theory is deeply influenced by Scotus, who holds that the soul is the substantial form of man, but also that the body has a form on its own.

Now, Fairley 1615 ascribes the theory of the unicity of the soul to Aristotle. The unicity of the soul is a necessary corollary of the immortality of the soul and of the Christian doctrine of the resurrection of the dead. In fact, in order to achieve the resurrection of the individual person, which is what the Christian religion claims, what makes a person an individual person must be regarded as immaterial and incorruptible in its entirety. Early Christians, in their attempt to establish an orthodox version of the notion of human soul, struggled against the 'Platonic' idea of a superindividual soul (for example, the notion of an agent intellect equally shared by all men, which found its way into Scholastic philosophy), which would not provide sufficient ground for the claim of the resurrection of individuals. Thus, Fairley seeks to ascribe the rejection of the plurality of souls to Aristotle himself, in order to gain his authority on the side of the Christian faith.

The eternity of the world and the analysis of the separability of the agent intellect are two Aristotelian doctrines that regents arguably over-interpreted in order to minimise disagreement with the Christian faith. The two cases present some differences: 1) with regard to the eternity of the world, Aristotle's doctrine is clear. The regents seem to offer an implausible interpretation in claiming that creation is not ruled out by the words of Aristotle. 2) With regard to the immortality of the human soul, *De Anima* III is not entirely convincing in proving it, and regents arguably carry out an interpretation *ex mente Aristotelis* when they claim that passages in *De Anima* III offer solid ground for the Christian doctrine.

Regents are still entirely within the Scholastic tradition in their attempts to find an interpretation of Aristotle which is coherent with the revelation. There is evidence to support the claim that the Humanist reformation in philosophy did not exert much influence in shaping the teaching in the Scottish universities in the first half of the seventeenth century for the following reasons: 1) Aristotle was still central in teaching, representing a uniform and coherent body of doctrines, whose pedagogical value was widely recognised; 2) until the 1650s the regents favoured the reading of Aristotle (among others) in the original language, as is proved by the number of Greek quotations in the theses; yet 3) Aristotle was still regarded as a 'Scholastic philosopher'; there is no evidence that regents abandoned the practice of commenting Aristotle in a Scholastic way; 4) when compared to the Scholasticism of the previous centuries, regents gave Aristotle an even greater role. I believe that this is a consequence of the Humanist reformation and of the separation of Scholastic philosophy from Scholastic theology; and also, a characteristic of Scottish Scholasticism in the seventeenth century. I shall qualify this claim in the next section.

2.2 *Aristoteles Reformatus*: a Reformed Scholastic aspect of the interpretation of Aristotle

In the seventeenth century, the fate of Aristotle's natural philosophy is linked to that of Scholastic natural philosophy. I argue that in the *Theses philosophicae* Aristotelianism and Scholasticism are two sides of the same coin. One attempt to assess the philosophical merits of Aristotelianism without reference to the traditional Scholastic reading of Aristotle was made later on in the century by an Aberdeen regent, George Skene, in his graduation theses entitled *Positiones aliquot philosophicae*, written for the class of 1688 at King's College. To my present knowledge, this set of theses is unique in contents. It is structured around an exposition of the main philosophical schools: Platonism, Stoicism, Epicureanism, Scepticism, peripatetic philosophy and Cartesian philosophy. What strikes the reader of these theses is the attention paid by the regent to the analysis of each philosophical school in its own right, thus offering what I believe to be the first work in the Scottish universities in the history of philosophy.

The section on peripatetic philosophy is of the utmost interest for the investigation of the interpretation of Aristotle. The section opens with the remark that:

> Philosophia peripatetica, magni quidem nominis olim, dum in scholis viguit *Stagyritae* authoritas, nil nunc nisi *Magni nominis umbra* est, quae subobscuris distinctionum involucris perplexa, anfractuosas rerum essentias intricatiores reddit. [V.1]

Skene is clearly under the influence of the enthusiastic endorsement of the 'new philosophy' which stemmed from Descartes' works and became increasingly important in Scotland after Descartes' first mention in Andrew Cant's *Theses philosophicae*, Marischal College, Aberdeen, 1654. The regent understands the philosophy of Aristotle to be essentially linked to the 'schools', and by his time is completely disregarded because of the excessive number of obscure distinctions in which it became involved. There is no doubt about the preference of the regent for the Cartesian philosophy, which alone is credited with the merit of providing a method for the acquisition of true knowledge (*ivi*, VI.1).

Despite the claimed identity between the philosophy of Aristotle and that of the schools, it is not entirely clear to what 'schools' Skene is referring. I believe that he has in mind Scholasticism in the form it took in the Scottish universities in the first half of the seventeenth century, since his idea of peripatetic philosophy coincides with the philosophy of the theses. In natural philosophy, two passages are revealing:

> 1) [prime matter] Non est *Pura potentia objectiva*, quicquid deblaterint *Thomistae*, realiter existit; ex re etenim non existente, nequit corpus componi. Non est *Forma*, nec in suo conceptu essentiali formam includit, licet naturaliter, absque omni forma existere nequeat. [V.11]
>
> 2) Accidens substantiae inhaeret, estque de ipsius essentia inhaerentia actualis. [V.6]

We have seen in part I that the regents hold the theory that prime matter is a metaphysical act, and that their reading of the Eucharist prompts the definition of accidents as essentially inhering in their natural substance. When ascribing these two theories to the 'schools' in general,[6] Skene cannot be referring to the Thomistic school (which rejects both claims), nor to the Scotistic school (which rejects the latter claim). I believe then that the peripatetic school as outlined by the regent concurs with the teaching of the theses in the form of a Scotistic Reformed Scholasticism. It is noteworthy that Skene ascribes to the 'Thomists' in general the doctrine of prime matter as 'pure objective potency', which is in fact a Scotistic notion.

Skene 1688 is not the only later set of theses which supports the direct implication between Aristotle and the claim that actual inherence is essential to the definition of accident. Robert Forbes, in his *Theses philosophicae*, King's College, Aberdeen, 1684, deals with the notion of accident in relation with Transubstantiation:

> *Accidentia realia*, quae divinitus existere possunt sine omni subjecto, comminiscuntur Doctores Pontificii, ad defendendam doctrinam suam de Transubstantiatione in Eucharistia: At nullum tale accidens reale admittit *Aristoteles*, cum ullus ex ejus germanis discipulis: illis enim (sicut et nobis) omnis forma materialis, sive *essentialis* sive *accidentalis* est *Modus subjecti, cui ita unitur et inest, ut impossibilis sit esse sine illo.* [XIX]

In 1684, the analysis of the Catholic doctrine of Transubstantiation does not differ from that of the first half of the century. Regents were very consistent with their criticism, which is based on the understanding of the definition of accident as mode of a subject which

[6] Regarding prime matter, Skene claims that prime matter 'really exists'. The regent seems to interpret earlier Scholasticism in the light of Cartesianism; yet, his remark that no composition is possible with something which does not exist echoes the Scotistic criticism of the Thomistic notion of prime matter as pure potency, which, as we have seen, plays a role in the attribution of a metaphysical act to prime matter.

cannot exist without its subject. I find three elements in Forbes's passage particularly relevant:

1) Forbes read the Catholic philosophers as holding that by divine power (*divinitus*) accidents can exist 'without any subject'. I argued in part I, chapter 4 that this theory seems not to have been held by any major Catholic Scholastics. I believe that regents misinterpreted, perhaps for polemical reasons, the Catholic notion of the aptitudinal inherence of accidents, which does not imply that accidents can exist 'without any substance', but only that accidents can, on a particular occasion, exist without 'their natural substance'. Catholic Scholastics and regents agreed on the traditional definition of accident as 'inhering in a subject', but disagreed on the notion of inherence: for the Catholic Scholastics it has to be aptitudinal, for the regents actual.

2) Forbes talks of 'accidentia realia', those accidents which can, according to the Catholic Scholastics, 'exist without their subject' and inhere in the substance of Christ. In the analysis of Transubstantiation, the formula 'accidentia realia' is not used in the theses of the first half of the seventeenth century. Neither is it common in late Scholastic philosophy.[7] Suárez does speak of what are now commonly called 'real accidents' in *DM*, 16, I, 3-4. While listing the types of accidents, Suárez claims that, among the accidents which affect their substance intrinsically, some accidents have their own entity and reality distinct from that of the substance and from that of other accidents; some other accidents are called 'modes': they are attached to other entities, and are really identical with their substance. Suárez gives the example of local presence as the same with substance, and figure as the same with quantity. The so-called 'real accidents' exert real formal causality, they are being by analogy, yet they are truly being.

Now, if an accident can be 'real' in the sense in Forbes's mind, it is not an accident as it is defined by Suárez. Suárez would not define accident as Forbes does, as 'a mode of the subject': all modes are accidents, but not all accidents are modes. Two developments seem to have occurred between the regents of the first half of the century and Forbes. First, the expression 'real accident' was used by Descartes, and Forbes is influenced by Descartes's use of this new expression. Secondly, again under the influence of Descartes, Forbes claims that all that affects a substance, and indeed all material forms, are modes: another use of the term 'mode' which is not the Scholastic one. I believe that both Scottish Scholastics and Scottish Cartesians arguably misinterpreted the Catholic Scholastic

[7] D. Des Chene, *Physiologia*, p. 113, claims that the notion of 'real accident' is based on the misinterpretation by Descartes and Boyle of reality for substantial existence. It seems that the 'reality' of real accidents is solely due to the real distinction between them and their substance, for example, in Suárez [*DM*, 16, I, 2], who accepts the Catholic account of Transubstantiation.

definition of accident: in fact, the later 'real accident' corresponds to the earlier 'accident which can exist without *any* substance'.

Despite these differences, Forbes's criticism of the notion of Transubstantiation is still the same as that of the first half of the century: the Scholastic notion of accident is contradictory.

3) Forbes writes that Aristotle does not accept the notion of an accident which exists without its substance, that is, a 'real accident'. Forbes is explicit in deriving the regents' criticism of Transubstantiation from the philosophy of Aristotle. I believe that the regents perceived their doctrine of the relation between substance and accident to be more faithful to the teaching of Aristotle than was the Catholic version.

I shall note that some regents do not invoke Aristotle's authority on every doctrine. There seems to be a profound difference between, for example, Forbes 1624 who ascribes the doctrine of the creation in time of the world to Aristotle, and King 1616, who writes that Aristotle is not concerned with the notion of prime matter (TP II.1) because he only admits matter as potential principle of compounds and not as a metaphysical act. As a consequence, Aristotle is not present in the analysis of prime matter, which is regarded by King as entirely Scholastic. This remark somehow balances those regents who ascribed Christian doctrines to Aristotle, respectively on creation and on the immortality of the soul.

Forbes 1684 and Skene 1688 shed light on the role of Aristotle in the graduation theses of the first half of the century. They tell us about how later regents understood the theory of their colleagues a few decades earlier: still in the 1680s Aristotle is perceived to hold a theory of substance which does not admit the notion of an accident existing without its substance, which is precisely the theory the regents criticised in the Catholic account of Transubstantiation. Thus, on this point, Aristotle is also perceived to be in agreement with a Reformed reading of the Eucharist. We can detect the traditional attempt to trace theories back to Aristotle, but the regents put forward a Reformed interpretation of Aristotle.

Can we say that regents were Aristotelian in their theory of substance? With regard to the notion of accident, they seem to be closer to Aristotle than their contemporary Catholic Scholastics were. Needless to say, the relation between substance and accident is not the only aspect which should be investigated before answering the question. The regents belong to the Scholastic tradition and there is no evidence for the claim that they sought to return to a historically accurate interpretation of Aristotle. Neither did they regard themselves as 'Aristotelians' tout court. In the Scottish universities in the seventeenth century the Humanist reformation of philosophy had left an identifiable mark on the attention paid to the Greek text of Aristotle rather than to the tradition of the commentaries on Aristotle. Yet, Scholasticism was still the main source for the interpretation of Aristotle, a Scholasticism which is influenced by Scotism and the Reformed reading of the Bible. Amidst many elements present in their interpretation of Aristotle (as we have seen in part I, chapter 3, section 2.3, regarding the unity of the compound) which qualify the regents as 'Scholastics', the regents directly ascribe the theory of accidents to Aristotle. We cannot say whether the regents in the first half of the century criticised the Catholic dogma of Transubstantiation explicitly on the basis of what was understood as a correct reading of Aristotle: they certainly did so in the 1680s, and ascribed this approach to their earlier colleagues as well.

In conclusion: 1) the regents tend to overlook the divergences between Aristotle and their philosophy, in the light of a Christian philosophy; 2) they also put forward an interpretation of the relationship between substance and accident which I also ascribe to their Reformed religion; 3) Aristotle was still regarded as a valuable source for philosophical enquiry, and it appears that the issue of a correct interpretation of Aristotle played an important part later on in the century, even if only in the afore-mentioned theory of accidents, a theory which has direct implications for the philosophical understanding of the regents' faith.

One final remark on the interpretation of Aristotle concerns John Seton, regent at Marischal College, and David Leech, regent at King's College, and the broader debate on the relationship between theology and philosophy which took place in Aberdeen in the 1630s and was finally halted by the depositions of regents following the National Covenant. Seton dedicates his 1631 *Theses philosophicae* to Aristotle, addressed in *Noncupatio*, page 1, as the 'Prince of philosophers' and a few lines after as 'our teacher':

> [...] Praeceptoris nostri ARISTOTELIS, laurea et palma, memoriaque sempiterna digni honori, rudem hanc tenuioris ingenii nostri opellam dicamus.

Despite the highest consideration for Aristotle, in 1627 Seton had made it clear that Aristotle's philosophy could not be regarded as anything more than a human enterprise:

> *Aristotelem*, quantumlibet acuto ac perspicaci valuerit ingenio, hominem tamen fuisse dicimus, a quo proinde nihil humanum alienum existimus oportet, humanum vero est interdum labi, ac errare, quicquid tamen ex propria sententia dixit, aliquomodo verum fuisse facile defendi potest. [TM 17]

How can we read this unique dedication to Aristotle in the context of the Aberdeen of the 1630s?[8] At the graduation ceremony of 1637 David Leech read before the audience an introductory oration titled *Philosophia Illachrymans* (Aberdeen 1637), an interesting source for the investigation of the cultural milieu of the Aberdeen colleges. Leech, among other things, claims that nowadays philosophy in Scotland is 'in tears', besieged by lack of material means, lack of cutting-edge research and, not secondarily, by attempts by theologians to impose their word in the philosophical domain. Seton's dedication to Aristotle and his awareness of Aristotle's fallibility may be regarded as, on the one side, a praise for philosophy in the person of the 'Prince of philosophers', but also as a hand outstretched towards theologians, in a period of heated theological debates. What I think is historically and philosophically central is that the regents go beyond a nominal praise for Aristotle and base their theory of substance on a reading of Aristotle which breaks with coeval Catholic Scholasticism and anticipates developments in early modern philosophy. The Aristotle of the regents was not the Aristotle of the Middle Ages.

3. Conclusions

In the first part of my work I have analysed the concept of prime matter. I decided to structure the exposition in the same way as Eustachius did in the *Summa philosophica quadripartita*: the reason being that Eustachius's approach has the advantage of clarity and completeness. The first three chapters focused respectively on the existence of prime matter, on its powers and on its properties. The fourth chapter dealt with the regents' reading of the Eucharist. The first three chapters form a unity in virtue of the harmony between the order of being and the order of exposition. In fact, the unfolding of the analysis from the definition of prime matter to its properties mirrors the metaphysical structure of prime matter, from the metaphysical act to the relation with form in the compound.

The first step is the evidence of the existence of prime matter and the definition of its concept (quod sit and quid sit): according to Aristotelian philosophy, a science cannot provide its own object, rather, a science expounds an object which is previously given to it. This way, answering the quod sit question is preliminary work to the analysis of prime matter. The second step is the investigation of the definition of prime matter as 'entitative

[8] I owe this contextualisation to Steven Reid.

act whose essence is being pure potency'. The third step is the analysis of prime matter as principle of natural compounds, that is, prime matter in its relation with form.

The unity between the first three chapters and the fourth one is justified by the following reasons: 1) the reading of the Eucharist is an instance of the logical-metaphysical problem of the definition of the accident and of its relation with substance; 2) the analysis of prime matter, quantity and extension that regents expound in their criticism of the Catholic theory of Transubstantiation sheds fundamental light on their general theory of substance.

The second part is about movement. I have chosen three topics which exemplify the debate in the theses: in the first chapter I have dealt with the general theory of movement, namely the meaning of nature, the act/potency theory, and the relation between movement and the categories. In the second and third chapters I have expounded respectively the theory of heavy and light bodies and the cosmology of the regents. The movements of heavy and light bodies and of the celestial bodies can only be understood in the light of the general Scholastic theory of movement, expounded in the first chapter.

The analysis of prime matter and movement are two parts of a unitary narrative. Prime matter is in fact the material principle of all natural bodies, bodies which are defined by 'being in movement'. The Scholastic notion of nature implies movement, and there is no movement without an inner principle of movement of the bodies, that is, nature. Furthermore, the theory of movement is only intelligible in the light of the theory of substance. Prime matter and movement together are the two central theories of the natural philosophy of the theses, even if they by no means are the whole of natural philosophy. They are historically important theories, since they are the background of the later reception of Cartesianism and Newtonianism.

I shall now present the conclusions of each chapter.

3.1 Part I: De materia prima

1) Prime matter is unanimously defined by the regents 'receptive entitative act'. The essence of this metaphysical act is 'being pure potency'. The notion of pure potency is traditionally employed by all Scholastics to define prime matter. Scholastics ground this definition in the Aristotelian theory of act and potency. A natural compound is the result of two principles which yield a unity per se: form, regarded as the actual principle, and matter, regarded as the potential principle. Regents claim that an unqualified 'pure potency' cannot be the component of a substance, since everything, in order to get in composition with something else, requires some sort of actuality. Therefore, they agree

with the notion of a metaphysical act proper to prime matter, aligning themselves with Duns Scotus against Thomas Aquinas. The natural philosophy of the regents thus bears the mark of Scotistic philosophy, which was indeed very influential in Scholasticism in the seventeenth century. Another aspect of the endorsement of Scotism is the metaphysics of essence. The regents claim that a substance exists in virtue of its essence, and therefore that there is no real distinction between existence and essence. The exposition of the quid sit question about prime matter has thus shown the general Scotistic approach of the theses.

2) The definition of prime matter as 'receptive entitative act' prompts the question of the powers of prime matter: that is, what prime matter is and does in virtue of its essence. The focus thus moved from the existence of prime matter to the analysis of its essence. The two powers of prime matter are 'being eductive' and 'being receptive'. This implies the relation with form: in fact, prime matter is essentially open to form and it cannot be understood independently of form. 'Being receptive' means that prime matter is a metaphysical act whose essence is receiving formal actuality. The regents once again engage with Thomistic philosophy, rejecting the principle according to which two acts cannot yield a unity per se. The problem is solved in Scotistic terms, by the distinction between metaphysical and formal act. Prime matter is 'metaphysically' actual, yet 'physically' pure potency. 'Being eductive' means that prime matter is the material principle of all forms (including the human soul), but more precisely, that material forms are educed from it. Material forms (forms without an independent existence from matter) originate from matter in virtue of an efficient cause, that is an agent triggering the eduction, and of the formal cause, the form which is triggered by the agent. Thus, the eduction of material forms and the information of matter by material forms are the same process, and the distinction is one of reason. In the conclusion of chapter 2 I have dealt with the set of theses by Dalrymple 1646, which breaks with traditional Scholasticism on the theories of the powers of prime matter and the real causality of secondary causes and seems to have been influenced by Descartes' theories.

3) According to the regents, the main property of prime matter is extension. In Scholastic philosophy, a 'property' is something that always and exclusively belongs to a substance (in Porphyry's words), without being included in its essence and in virtue of its essence. Prime matter is not defined as 'something extended', yet, according to the regents, it is necessarily extended. The regents always attribute to matter 'extension in place', thus breaking with their Catholic colleagues, who claim that only 'extension in ordine ad se' can be considered a proprium of matter. The distinction is important: in fact, the regents hold that prime matter is spatially extended before the information by form, and spatial extension must be predicated of matter regardless of its being part of a compound or not.

Prime matter is also the subject of accidents, so that not all the accidents of a compound depend on form. Even if the regents stress the notion of the unity per se of the compounds, their theory of extended matter and the Scotistic notion of bodily form may lead to a form of dualism within the compound. It is then arguable that the Scotistic Scholasticism of the regents proves itself to be close to later developments of early modern philosophy, especially Descartes'. The theory of extended matter leads us to the discussion of the Eucharist.

4) The rejection of the Catholic account of Transubstantiation helps us to clarify the theory of substance of the regents and to identify Reformed elements in their Scholasticism. The Eucharist is a theological notion, which nonetheless bears consequences in philosophy, since the Catholics developed philosophical doctrines in order to account for the supposed transubstantiation of the substance of bread and wine respectively into the body and blood of Christ, and of the preservation of the accidents of bread and wine throughout the process. As Reformed philosophers, the regents did not accept the interpretation of the Last Supper as a miracle, but saw it as a symbol. What is central to my scope is that the regents refuse to engage with theology and that they respond to Catholic Scholastics by deploying arguments which profoundly shape their philosophy. The regents claim that the traditional definition of accident, already expressed in the works of Aristotle and Porphyry, excludes the possibility of an accident existing without its substance. Just as prime matter is always extended in place, an accident always inheres in its substance. The form of the argument is the same in both cases, since quantity is regarded as an accident of matter: there is no real distinction between an accident and its actual inherence, therefore, an accident cannot be separate from its substance. I believe that this theory is a characteristic of the theses, and that it is the product of both the reading of Aristotle and the Reformed religion of the regents. They seem to exploit the Scotistic notion of formal/modal distinction in order to account for the necessary unity within the same compound of matter, quantity and place: these elements are defined in different ways, yet they are always conjoined. It is arguable that regents downplayed the importance of the real distinction in favour of the modal distinction.

3.2 Part II: Movement

1) The first chapter of part II is about the general features of movement and lays out the framework for the analysis of the movement of heavy and light bodies and celestial bodies. The regents held a Scholastic theory of movement: movement is described as the process

from a terminus a quo to a terminus ad quem, which means, from an old form to a new one. Movement is always referred to as regarding form, rather than the whole compound. The commonly accepted formula in the theses is that movement is a tendency, a way and a flux of form from one terminus to another. In this regard, the regents do not adopt the Scotistic theory of the *forma fluens*. In general, the regents' theory of movement seems to have been less influenced by Scotism than their metaphysics. An important question is that of the relation between movement and the categories. The regents held that generation (that is, the process from the absence of a substance to a substance) cannot be properly called 'movement' because it takes place in an instant, while movement necessarily takes place in time. Furthermore, a movement is between two contraries within the same species, and a non-substance and a substance are contradictories: therefore, generation is not a movement. The standard theory of the theses is that movement falls in the categories of quality, quantity and place: respectively, alteration, augmentation/diminution and local movement. Some regents raise the question of whether the categories of quality and quantity should be excluded from the number of movements. When the answer is affirmative, the reason is the same as for generation: movement only occurs in a succession of time.

2) A fundamental part of the Scholastic theory of movement is the theory of natural places. Bodies tend towards their respective natural places, in virtue of the proportion of the elements they are composed of. Thus, bodies with a predominance of earth or water will be 'heavy bodies' and will fall towards the centre of the universe; similarly, bodies with a predominance of air or fire will be 'light elements', and will move upwards towards the sphere of the moon, the upper limit of the sublunary world. The analysis of this theory calls into question the historical debate on the coherence between the *Physics* and the *De generatione et corruptione* of Aristotle. In fact, the regents seem to favour the terminology of the *Physics*, since they account for the movement of heavy and light bodies in terms of their form. Heavy and light elements are included in the behaviour of a body dictated by its form (which, according to the regents, is the same as the nature). Heaviness and lightness are absolute (that is, non-relative) concepts: something is heavy or light in virtue of its nature, not in relation to something else. The natural end of a movement is rest: and this is true for heavy and light bodies as well. Reid 1626 seems to accept exceptions to this principle, when he claims that, for example, fire moves in order to move, not in order to rest. It is arguable that this theory hints at a break with the Scholastic tradition. Another key element is the notion of the 'mover' of heavy and light bodies: regents claim that heavy and light bodies move downwards and upwards in virtue of their forms. One final aspect concerns final causality: in Scholastic natural philosophy, final causality is the type

of causality exerted by the end of a movement. The regents accept this notion, and I argued that they offer compelling arguments in favour of the distinction between efficient and final causality.

3) The cosmology of the theses is still based on the assumption of the difference in nature between sublunar and celestial bodies. The difference lies in the fact that sublunar bodies are subject to corruption, while celestial bodies are not. I argued that the paradigm of the difference in nature between sublunar and celestial bodies is a good example of the general style of Scholastic philosophy. I employed the term 'paradigm' because it seems to account well for the complexity of a worldview which is not based on empirical evidence, but rather justifies the empirical evidence brought in its support. Under this point of view, the Scholasticism of the theses appears to be still traditional, despite an increasing interest in the mathematical analysis of the heavens, as the *Theses astronomicae* show. One Scotistic element is the acceptance (at least as a logical possibility) of a natural void: the regents claim that a movement in a void, if such void exists, is possible, and it would take place in a succession of time. As a consequence, regents hold that the Aristotelian position of the infinite speed of a body moving in a void is wrong, since it exclusively relies on the notion of the resistance of the medium, without acknowledging the internal resistance of bodies. The conclusion of chapter 3 raises the question of the interpretation of book VIII of the *Physics*, and of the role of natural theology in the theses. I believe that the regents, influenced once again by their faith, exclude natural theology from the theses, by limiting the traditional principle 'omne quod movetur ab alio movetur' to natural philosophy.

3.3 Final remarks

I have defined the natural philosophy of the *Theses philosophicae* as 'Eclectic Scotistic Reformed Scholasticism' for the following reasons:

1) It is a form of Scholasticism. The regents are still much indebted to the traditional philosophy of the schools, in terms of contents, form, references and structure of the exposition. If it is true that they abandoned the Medieval practice of the quaestiones, they nonetheless wrote in the style of the Scholastic textbooks of the early modern period. Neither can their philosophy be called 'Aristotelian' tout court. Even if Aristotle is the main source of inspiration, the regents do not generally seem to agree with the Humanist agenda of interpreting Aristotle outside the Scholastic framework. Yet, it seems that the regents benefited to some extent from reading Aristotle in the original text: for example, their theory of substance has some decisive Aristotelian features.

2) It is a Scotistic natural philosophy: the metaphysics of essence, the theory of prime matter, the use of the formal/modal distinction, the notion of bodily form, the theory of the movement in a void and the theory of the void as a geometrical space potentially occupied by bodies drew inspiration from the philosophy of Duns Scotus. The regents can be said to be part of the long tradition of Scotistic philosophy.

3) It is, nonetheless, an eclectic natural philosophy. The regents never openly claim to philosophize in accordance with the principles of the philosophy of Scotus, in a period in which, in the Catholic world, the division in philosophical schools was very strong. Alongside this aspect, there is evidence that the regents were keen on learning from Catholic Scholastics, and make use of their philosophy in a creative and original way. Even if regents rarely stand out amidst their colleagues for particularly personal theories, every set of graduation theses shows a peculiar character. Graduation theses can be successfully treated as a 'school' within Scholasticism, even if they do not show an unequivocal uniformity.

4) It is a Reformed natural philosophy. One fundamental principle of unity among the theses is the Reformed religion of the regents. Philosophical debates are posterior to the acceptance of the Christian faith in its Scottish Reformed form. Together with the traditional principles of Scholasticism in its attempt to harmonise revelation and human reason, the *Theses philosophicae* put forward two theories which I openly ascribe to the Reformed faith of the regents: 1) the actual inherence of an accident in its natural substance as part of the definition of the accident; and 2) the rejection of natural theology as an object of philosophical investigation.

In the first half of the seventeenth century Scholasticism was still the traditional philosophy of the Scottish universities. It was a lively and much-debated common philosophy, which appears to have shared roots in Scotism and Reformed religion, even if it was not as internally coherent as a school can be. The regents were highly acquainted with contemporary Catholic Scholasticism, to such an extent that it is proven that Scholasticism in Scotland did not end with the Reformation. Scottish Scholasticism greatly benefited from the Reformation and to a certain extent from Humanism: I believe that some theories in metaphysics and natural philosophy, such as the theories of substance and prime matter, prove the constitutive openness towards later developments of early modern philosophy and the degree of originality of Scottish Scholasticism. It is then arguable that the influence of Scholasticism in Scotland extended well beyond the reception of Descartes in the early second half of the century.

I also believe that the investigation of Scottish Scholasticism can shed light on the still underexplored field of Protestant Scholasticism, and decisively influence our reading of the philosophical revolution of the seventeenth century. If we accept my arguments, it appears that Scholastic philosophy in Scotland was not simply a reaction against Catholic Scholasticism, or a heritage of the pre-Reformation curriculum of the universities. These aspects are certainly part of the narrative, but innovation and reinterpretation of Scholasticism are as well. Scottish Scholasticism seems to anticipate early modern philosophy, arguably in virtue of its Reformed character. I am convinced that three distinct directions of research shall complete the analysis of Scottish Scholasticism: 1) its relation with the general cultural life of Scotland in the seventeenth century; 2) its relation with Scholasticism in other branches of philosophy; and, finally, 3) its relation with modern philosophy and the Scottish Enlightenment.

Appendices

1. *Theses physicae*, G. Robertson, 1596

Complete title: *Theses philosophicae.* Publicly discussed on August 2nd at King's College of Edinburgh University, at 8am, as we are informed on the title page. Printed by Henry Charter in 1596, in Edinburgh.

The theses are divided into four sections: logic, physics, *Theses sphaericae* on astronomy and ethics, for a total of 16 pages.

This is the first extant set of graduation theses for the Faculty of Arts of the Scottish universities. The practice of printing graduation theses had been recently introduced in Edinburgh. The first graduation theses of the Faculty of Arts followed two theological theses, in 1594 and 1595, which are the earliest theses in Scotland. Andrew Melville and the printer Robert Rollock are the founders of the printed theses in Scotland.

Despite the novelty of the format, the early date and the brevity of the work, Robertson's theses are very similar in content and structure to later theses. It seems that the establishing of the practice of publishing graduation theses gradually led to more complex and longer theses (in Edinburgh especially in the 1610s and 1620s) but did not bear consequences in the curriculum being taught.

Thesis I. Because of the lamentable original fall, not only by the paralysis of a dissolute affect, throughout all its acts, is will darkened, but mind as well, by Theban sphinxes and Cymmerian obscurities.
The grievous human condition is in need not only of the cure of practical philosophy, but also of the collyrium and sun of contemplative science.

2. In this worldly machine, the highest maker refined the whole so ingeniously, and connected the superior beings to the inferior with indestructible ties.
Physical science is extremely necessary and useful.

3. The subject of Physics is the natural body as endowed with a nature. We establish a way of considering it as natural, just as common opinion does.

Although the arts consider the natural body as subject of their operation, they do not admit the way of considering it as flowing from the nature of the (same) subject: the arts obtain the (proper) mode of the subject, from the point of view of an end which is only ours.

4. Prime matter is a substantial, bodily and perishable being.

We are not afraid to ascribe existence to prime matter alone considered without form, although an imperfect one, which the incoming act of form makes perfect.

5. Potency and quantity follow from the nature of prime matter.

We do not oppose the opinion of those who assert that the succession [flux] *of all accidents is from a form or an agent.*

6. The mass [moles] of prime matter is unpolished and disordered, and the incoming form polishes and orders it.

1. Before the arrival of form accidents which depend on matter are interminate.

2. Thus, the quantity of prime matter is interminate.

7. Matter is the subject of inhesion [inhaesionis], form is the condition of inhesion [inhaesionis] of the accidents (excluding the spiritual accidents), and inherence completes [terminat] the accidents.

It is not impossible that some accidents remain the same in begotten and corrupted being, because of the change of the condition of inhesion and of the termini.

8. The potency of prime matter (just like prime matter, which remains the same in itself), is made fit and arranged towards several more noble forms, by the change [accessione] of certain conditions.

Therefore, we believe that prime matter does not differ from second matter, that first potency does not differ from second potency, and that they truly differ as more perfect from less perfect, and as absolute from modified.

9. In order for form to come from the potency of matter, it is required not only that matter is capable of receiving form, but also that form depends on matter in three regards: in production, in being and in operation.

1. Therefore, the forms of inanimate beings in particular, namely mixed beings and elements, must be considered as coming from the potency of matter.

2. In some sense, the sensitive and the vegetative souls come from the potency of matter (although they are in some measure raised by matter's condition, as it is proven in the increment [accretio]*), and truly the rational soul does not come from the potency of matter at all, if not perhaps from the point of view of* [it being received by] *the capability of matter.*

10. Matter remains the same in begotten and corrupted beings.

Thus, form is the whole quiddity and essence of a thing, matter instead is its vehicle [vehiculum].

11. The matter of the heavens is actuated in itself.

1. And so, there is no difference between the heavens and their nature.

2. The nature of the heavens cannot be the middle term in the demonstration of the movement of the heavens: in fact, middle term and subject would not be different.

12. The intelligence is the middle term in the demonstration of the movement of the heavens.

The middle term in the demonstration of the movement of the heavens is an external cause, in respect of information: unless we believe that assistance is an internal cause.

13. Generation and corruption refer to [determinant] one single mutation.

These two terms only reflect two termini.

14. The matter of the heavens is different from the matter of perishable bodies: in fact, the former is not in potency towards form, while the latter is never devoid of potency.

Although action and passion occur between celestial and perishable bodies, no reaction and repassion [repassio][1] *occur at all.*

15. The species of the elements is different from that of the mixed bodies.

The forms of the elements, with regard to excellence, do not remain in mixed bodies.

---

[1] "*Dicitur repassio, qua agens vicissim patitur ab eo in quod agit; seu, receptio effectus ab agente imbecilliori: v. g. reception frigoris in ferro candente, ab aqua, cui immergitur.*" É. Chauvin, *Lexicon rationale, sive thesaurus philosophicus*, Rotterdam 1692, art. Repassio.

16. It became common among everybody to believe that contrary beings cannot inhere in the same body, in their normal conditions.
The qualities of the elements in mixed bodies are shattered and restrained [fractae et castigatae].

17. According to Aristotle, an increment requires three conditions in order to occur, 1. that the increase [accessio] of a quantified body takes place, 2. that the body whose increase takes place is augmented in respect of its minimal parts, 3. that the body remains numerically the same.
To preserve the true sense of an increment, we can ascribe it only to animate beings, and to beings becoming ripe and growing up: in fact, when they are ripe and adults, a pause in the increment occurs, because of both the satisfied intention of nature and the disposition of matter.

18. Every Physical form enjoys this great privilege: to inform and actuate matter and any of its individual parts.
The soul is in the whole body and in all of its individual parts.

19. Despite being devoid of quantity, form takes on quantity from matter by accident, since it extends to match the extension of matter.
Therefore, although the soul with regards to its real being is in the singular parts of the body, with regards to quantitative extension, taken on from matter, it is not in the parts considered alone, but it truly expands itself throughout the whole organic body.

20. The effects which emanate from their causes cannot be separated from the site of their causes.
When the souls, which are in the individual parts of the body, emanate their effects, the faculties originally spring up in every part of the body.

21. The organic faculties of the soul have designated and determined organs which serve them in their operations.
Therefore, whichever faculty of the soul is not subjectively in any part of the body; however it is in it originally.

22. It is familiar to all who just moved the first steps in philosophy that the nature of the genus is included in the nature of the species.

We do not agree with those who believe that the vegetative soul is the specific form of a plant, and that the sensitive soul is that of an animal: in fact we argue for silent and hidden forms in plants and animals.

2. *Theses physicae*, A. Aedie, 1616

Complete title: *Theses generales, logicae, ethicae, physicae, sphaericae*. Printed in 1616 in Edinburgh by Andrew Hart. The copy in Aberdeen university library I read lacks the title page, which usually informs on the place and date of the public graduation ceremony.

This set of theses of Marischal College is the first one available for Aberdeen. It was printed in Edinburgh since no printer was working in Aberdeen at that time. All later Aberdeen graduation theses were printed in Aberdeen by Edward Raban.

It is divided in five sections, the first being a section on 'general theses' on the relationship between philosophy and theology and on the order of philosophical disciplines. What is remarkable about these theses is the focus on special physics, which gives us an idea of contents of the curriculum which are usually missing from other theses. Alongside matter, increment and the nature of the soul, the regent expounds his theories on natural monsters, rainbows, colours, and odours. This gives a distinctive encyclopaedic flavour to the theses, enriched by quotations of classical authors, the reference to rare and imaginary animals and plants, and the use of Greek.

Another uncommon yet revealing feature is the listing of *Problemata* which ends every thesis. In these parts, the regent raises some questions, whose answers are either '*affirmatur*' or '*negatur*' (or '*affirmo*' and '*nego*') or '*distinguitur*' (or '*distinguo*'). Theses questions might be examples of the kind of topics students had to discuss in order to give proof of their preparation and rhetorical skills.

Thesis I

The appetite of matter is defined by the Philosopher in I Acroa. text. 81 as the natural propension of matter towards forms indistinctly.

Appendix I. It is necessary that when one form leaves matter another one arrives.

2. As long as matter is determined towards a certain form, it has potency and appetite towards it closely and intensively.

3. No form can satisfy the appetite of matter extensively.

4. While matter has one form so perfect in act that it cannot receive a nobler one, nonetheless it desires another one.
5. Hence matter is called the cause of preservation. 2. de Ort. ch. 9 since the totality of sublunary forms subsists in a continuous series by the introduction of recent forms.
6. The same matter is also considered the cause of the corruption of things. 9. Metaph. ch. 9. since, while matter admits the qualities which drive out the previous forms, it also plots for the corruption of the compound.
7. In fact, matter is the principle of being as much as of non-being of perishable things. 7. Metaph. ch. 7.
8. It is not only the cause why generation is possible at all; but also the cause why the vicissitude of generation and corruption can be perpetual.
9. The ancient philosophers who made up the story that prime matter is God made a miserable mistake, because God's nature, being the purest act, is as far as possible from matter.
Problem I. Whether matter is common to all bodies, or is proper to each body in its species. Both true.
2. Whether the matter of the heavens and of the inferior bodies is the same and not the same. True.
3. Whether the matter of contraries is the same and not the same. True.
4. Whether the matter of all the elements is one, despite being the fourfold matter of the elements. Both true.
5. Whether form (since it is coeval to matter), rather than matter, is the cause of corruption. Distinction.

Thes. II. A monster is a living natural body provided with a certain defect of nature.
Appendix I. Hence 2. Phys. Acroas. ch. 8. A monster is not inappropriately called ἁμαρτήμα τής φύσεως[1] [mistake of nature].
2. And it is not wrong to distinguish between a remiss or intense degree of nature and a wandering off of nature.
3. They are insane those who exclude the females from the number of humans (forgetting that their mother was a female), and put them among the monsters.
4. Neither do we agree with those who considered and still consider the pygmies and the giants, the dwarves or the little boys to be monsters.

[1] I reproduce the Greek text as it appears in the text. The only change is the adoption of our contemporary style for the characters. All translations from Greek are mine.

5. Neither do we agree with Martin Veynrich, who includes those with six fingers and the cyclops or monocules among the monsters.

Problem I. Whether all monstrous births from humans must be counted as human. False.

2. Whether all monsters among humans must be baptised. False.

3. Whether all monsters must be killed immediately after birth. Distinction.

4. Whether mermaids and centaurs are only figments, or they also exist in reality. Former false, latter true.

5. Whether there are monsters among plants (such as the vegetable lamb of Tartary), as there are among animals. True.

6. Whether any monsters existed before the fall. False.

7. Whether the judgment by Augustine in the Enchiridion, ch. 87. is true: that human monstrous bodies in this life will be given their integrity and perfection back in the last resurrection. True.

Thes. III. Ο Θέος καὶ ἡ φύσις οῦδεν ματην οῦδε αλόγως ποιοῦσι. [God and nature do nothing in vain or without reason] I de Coelo, ch. 5 and 2 de Coelo, ch. 11.[2]

Appendix I. It seems that [God and nature] arranged the lines of hands, forehead and of the whole bodily mass, provided that it is externally different, according to an end.

2. So that we do claim that Physiognomy, Metaposcopy, and Chiromancy are in things produced by GOD and nature.

3. And that these arts are called conjectural; [name] that is added to them with respect to the practitioners rather than with respect to the things they deal with.

4. They are wrongly considered as magical and forbidden arts.

5. Ignorant people wrongly condemn the supporters of these practices as unworthy of the Christian community.

Problem I. Whether Chiromancy can be proved from the evidence of the Scripture. True.

2. Whether Aristotle said correctly in I De Hist. animal. ch. 15 that it is possible to judge on the length of a life from the length of the lines of the hand. True.

3. Whether signs of a violent death can be gathered from those marks which are commonly called divine characters. True.

4. Whether different conclusions about the death can be conjectured from different marks. True.

---

[2] A standard text of the *De coelo* reads: ἡ δὲ φύσις οὐδὲν ἀλόγως οὐδὲ μάτην ποιεῖ, book 2, chapter 11, 291 b 13-14 (Karl Prantl, Lipsia 1881) It appears that the regent has added the term ‘god’, missing from Aristotle’s passage. The reference to book 1, chapter 5 is probably wrong.

5. Whether it is possible to determine something certain from lines inspected properly or casually without the observation of circumstances (what Egyptian vagabonds do). False.
6. Whether it is possible to infer from external signs the very virtues of the souls or rather the propensities towards some virtues or others. The latter is true.
7. Whether the same signs in different (as are commonly called) heaps [montibus] can mean different things. True.

Thes. IV. Increment is defined as a movement tending from imperfect quantity to perfect quantity, by the conversion of nourishment into substance with the loss of a greater imperfection; therefore, all the parts of the body, together with the form, are made proportionally bigger in order for the living body to carry on the functions proper to life, once it has gained the right magnitude. I. de Ort. text. 25. and 31. and 35. 2 De Anima. text. 14.
Appendix I. The body is augmented not as a whole but as a potency, because of the matter from which the accretion of the body results.
2. An element cannot be the proper nourishment thanks to which a living body can grow.
3. Indeed neither gold, which the Chemists call drinkable, since in the end it cannot be made similar [to the substance]. Scal. exercitat. 201.
4. Neither can tobacco [Nicosiana illa ludica], whatever the Tobacconists babble to the contrary.
5. In an increment, the parts of a living body do not grow according to the form of the part.
6. Therefore, Aristotle refers to the form of the whole. I. de Ort. ch. 5. when he writes that the parts are increased in respect of their form.
7. Therefore flesh grows by an inch as flesh, not as this signate flesh.
Problem I. Whether by accretion the subject remains the same according to the material aspect, or instead according to the formal one. The latter is true.
2. Whether a living body can live up to and longer than one year without nourishment. True.
3. Whether the soapwort plant, granted that it devours iron, also digests it and converts it into the substance of the body, in order to grow. False.
4. Whether the Chameleon only feeds on air and odours, as Pliny and others claim. False.
5. Whether the herrings grow and feed on water only. Distinction.

6. Whether serpents eat only earth. Experience denies it. Neither is this opposed to Genesis ch. 3, v. 14.[3]

7. Whether God three hundred or four hundred years after his birth can be accounted for a true and properly called increment. False.

Thes. V. The rainbow as in chapters 4-5. of 3. Meteor. can be defined as an arch in a bedewed and hollowed cloud, which shows the various species of colour of the different parts because of the opposite sunrays, or of the refraction of the moon.

App. I. The solar rainbow does not meet our vision unless we are between the sun and the cloud thanks to which the rainbow glitters. 3. Meteor. ch. 4.

2. In our climate, a rainbow cannot be seen towards South, since it never occurs that we are between a cloud and the sun in that direction.

3. A rainbow is usually smaller than a semicircle, and it is never bigger. 3. Meteor. ch. 5.

4. We do not agree with Mirandula, who (in book 2. ch. De Humanae studio Philosophiae) claims to have seen a rainbow in a complete or almost complete circle.

5. Were a rainbow in a complete circle, it would follow that a straight line would pass right through the centre of the rainbow, of the sun and of the horizon or through the eye of the observer on ground.

6. Since the sun stretches more at sunrise or at sunset, the arch of the rainbow is bigger; on the contrary, the rainbow is smaller when the sun is at its highest on the horizon.

7. The biggest rainbow of an entire day takes place when the sun is either rising or setting, and the smallest takes place in the remaining moments, when the sun is at noon or does not appear at distance.

8. The rainbow can some times appear with a full arch and some other times with a broken arch, if the cloud which enables the impression of the rainbow is divided, like one part in the east and the other in the west.

9. Two solar rainbows are frequent, three are rarer: the first is due to the reflection of the solar rays, the second is due to the first, the third to the second, with a clear inversion of the colours.

10. Lunar and solar rainbows appear constantly, but the former is rarer and, if we are to believe the Philosopher, the latter is most rare during autumn.

---

[3] *"And the LORD God said unto the serpent, Because thou hast done this, thou art cursed above all cattle, and above every beast of the field; upon thy belly shalt thou go, and dust shalt thou eat all the days of thy life." King James Bible*, Book of Genesis, 3:14.

11. In a lunar rainbow only white colour with somewhat blackish lines is discerned, because the weakness of the lunar rays can hardly penetrate the darkness of the night and the thickness of the clouds.

Probl. I. Whether rainbow existed before the flood or not. True.

2. Whether a solar rainbow sometimes appears all white, as Melichius refers while commenting on the second book of Plyny. False.

3. Whether Scaliger in exercitat. 80. sect. I. correctly claims that three colours usually appear in the rainbow, due to the variety of the matter of earth, water and air existing in the cloud. True.

4. Whether a fourth colour can be added to the three colours of the rainbow, ξάνθος [yellow] or golden, which is due to the mixture of scarlet and green, according to the Philosopher 3. Meteor. 5 and Scal. in the passage mentioned above. True.

5. Whether Cardan correctly calls the rainbows a pure figment of the eye. False.

6. Whether a rainbow can be visible on the surface of the sea (as some claim), where there is no dewy cloud. False.

7. Whether a rainbow always anticipates a future rain. False.

8. Whether Seneca, book I natural. quaest. ch. 6, correctly claims that a rainbow in the east is sign of rain, in the west of nice weather. Whether more correctly, in book 2. natural. hist. ch. 6. Plyny claims that neither rain nor nice weather can be predicted with confidence from a rainbow. Former is false, latter is true.

9. Whether a rainbow naturally or rather above or beyond [supra/praeter] nature indicates a non-future inundation of earth. Only the latter is true.

10. Whether rainbows are more common in the East than in the West. It is probably so by natural causes.

11. Whether in a solar rainbow an entire image of the sun can be represented, as in the parhelions.

12. Whether a lunar rainbow never occurs unless in decima quarta or decima quinta moon, or around this time, as the Philosopher claims in 3. Meteor. ch. 5. True.[4]

13. Whether the theologians correctly claim that in a solar rainbow there are two colours which are especially visible: internal blue and external scarlet. The former stands for the destruction of the world in the flood, the latter for its eventual destruction by fire.[5] This does not seem either inconsistent or impious.

---

[4] Traditionally, this is the time for the celebration of Easter, in the third week of the first lunar month, around the spring equinox.

[5] *"Secundum fidem Christianam* [...] *cum Apostolo credimus coelum et elementa omnia igna purganda"*, Reid 1614, TP 22.6.

Thes. VI. Experience shows, and the Philosopher confirms that 2 de part. animal. ch. 7, that the brain of man is by nature colder and moister than that of other animals.

App. I. Hence it is evident why man is surpassed by many animals in the sense of odour or olfaction. 2. de Anima, ch. 9. text. 92.

2. And also why the odours which are well perceived are those which are more excellent in the extremes and not those which are more remitted in the middle.

3. And why little or nothing is smelled in winter rather than in summer, or in great heat or cold.

4. And why the latter [odours] are more strongly affected than the former.

5. Thus, the upbringing and habits can change a lot the mixture of the brain, as it appears in those subjects who live in prisons or in squalid places.

6. Although men are surpassed by beasts regarding the excellence of perception, on the contrary men surpass beasts by far in the eminence of judgment.

7. Only man has this sense with regards to its perfection. Scal. exercit. 247.

8. The theory of Bodin, in book 4. Theatri Naturae. is praiseworthy. He claims that the wisdom of the Maker is great, since if he had given a sharp and accurate olfaction to men, they would have not been able to bear not only other people's smell, but not even their own.

Probl. I. Whether odours feed, granted that they restore. False.

2. Whether materially, or only formally, as the objects of the remaining senses are perceived. Only the latter is true.

3. Whether man alone among all the animals receives pleasure from odours. Distinction.

4. Whether odours can sustain life for some time, as Plyny writes about Democritus. True.

5. Whether the dogs which the Scots use to follow the tracks of men have a different mixture of the brain from the others, and hence a different way of smelling, both are true.

6. Whether the Astomi people that Plyny speaks of in book 7 of naturalis historiae, ch. 2 can live off odours alone. False.

7. Whether tomatoes [mala aurea] and the very genus 'pomes' are diminished by the mere emission of odours or of steams and vapours as well. Only the latter is true.

Thes. VII. The Philosopher. I de Anima, ch. I. and 4. Then also in ch. 5, book 3. claims that the soul is χωρίστην απὸ του σώματος [separate from the body], and in ch. I, book 2. de Anima calls it μόρφην, and defines it as such.

App. I. It appears that the Philosopher was aware of and approved of the immortality of the soul.

2. As a soul separate from body requires, it is also necessary a body just as matter which form informs, since form without matter cannot really be.
3. Therefore we claim that the resurrection of the dead is probable by reasons in a certain way Philosophical.
4. Indeed, it does not follow from what has been said that the soul does not die once separate from the body, since it seems that, from a Physical point of view, anything which is generated dies, like the day. I de Coelo, last chapter.
5. And since the soul goes back to its matter as form, it is not possible to accept the μετεμψύχωσα [metempychosis] of the souls in the fashion of the Pythagoreans. 2. De anima. end of ch. 3.
6. Those belonging to the herd of pigs of Epicurus[6] wrongly claimed that souls die with the body.
7. Neither Origen's mistake seems acceptable to ethnicists and philosophers, since it postulates infinite souls created in the beginning, which by whatever case or chance are placed in a body.
Probl. I. Whether the Physical form is the soul or otherwise. False.
2. Whether the soul is whole in the whole body, and whole in singular parts, it can be discussed pro and against.
3. Whether the sensitive soul of beasts and men, and the vegetative soul of plants are of a common and same substance. False.
4. Whether in what Arist. said in 2. de hist. animal. ch. 3. ψύχεν θυραθέν επεισί ειναι [some claim that the soul is external], he recognized that the soul is by inspiration; it seems probable.
5. Whether the rational soul acts in the body without bodily organs, as when it is separate. True.

[6] Horace, *Epistulae*, I, 4, 10.

3. *Theses physicae*, J. Reid, 1626

Complete title: *Theses philosophicae*. Discussed on July 31st 1626 ('*ad diem* Pridie Calend. AUGUSTI') in Edinburgh. Printed by John Weittoun in 1626 in Edinburgh.

The set of theses is divided into five sections, on 'general theses on disciplines', logic, ethics, physics and *Theses sphaericae* on astronomy.

James Reid is one of the most important regents of Edinburgh University in the first half of the seventeenth century. He is the author of five sets of theses (1610, 1614, 1618, 1622, 1626), among the longest and most detailed in all Scottish universities. The 1626 set is his last academic work and stands away from both his previous sets and the rest of the theses for his unique theory of movement, which seems to break away from the Scholastic theory that every movement is always directed towards an end [which I analyse in part II, chapter 2, section 4].

Reid's theses enable us to investigate the development of the philosophy curriculum as it was taught by the same regent over twenty years. This is possible in particular for Edinburgh University, where from 1610 to 1628 three regents, Reid, King and Fairley held the position for almost two decades.

Thesis I. *By unanimous consensus of philosophers, matter in begotten and corrupted bodies is numerically one and the same.*

1. In the succession of generable and corruptible things, also matter which is numerically one can be under forms distinct by genus.

2. In fact, no form actually gives number [numerical identity] to matter, either considered according to its own entity, or related to the nature of things.

3. Since numerical and real unity cannot be without existence, the prime matter of all things will also have its own existence not depending on form.

4. Although one and the same portion of matter can receive several forms in succession, it will not change several existences because of this; on the contrary, its existence must be said to be one and the same under any forms.

5. Since in fact existence is only a mode of a thing, not really distinct from it, and since some things are complete, some others are incomplete; so, also one existence can be said to be incomplete, and another one to be complete.
6. Accordingly, the existence of matter and form is partial and incomplete: that of the compound is complete and total.
7. There is an existence in things, distinct from the formal existence, which is only proper to the compound.
8. Thus, the existence of form and the formal existence are different.
9. And it is very different to exist in nature simply and completely, and to exist formally.

II. Βαρύτατον τὸ πᾶσιν ὑφισάμενον τοῖς κάτω, κουφότατον δὲ τὸ πᾶσιν ἐπιπολάξον τοῖς ἄνω φερομένοις. *I. de Coel. 3.*[1]
1. Α'πλῶς κοῦφον λέγομεν τὸ ἄνω φερόμενον καὶ πρὸς τὸ ἔσχατον, βαρὺ δὲ τὸ κάτω καὶ πρὸς τὸ μέσον. 4. de Coel. I.[2]
2. Thus, it is only natural to earth to move towards the lowest place: and to fire to move towards the highest one, and natural to both to remain in them. φέρεται φύσει καὶ μένειν ἐν τοῖς οἰκείοσις τόποις ἕχασον τῶν σομάτων. 4. Phys. 4. text. 30.[3]
3. Earth does not have a bigger propension to rest in the lowest place, than fire to movement, in the highest.
4. As much as movement towards the sphere is given to fire beyond its own nature, why can't we similarly believe that rest is assigned to earth, beyond its own nature?
5. Nature is the principle and cause only of the movement of earth, not of its rest.
6. And by its nature and natural condition, earth is mobile and not immobile.
7. Therefore, since every rest, also the one following from natural movement, is privation of the same movement: so, nature does not specifically seek rest, neither it strives for [intendit] it.
8. From this we conclude, first: that it is very different for a thing to remain in its own natural place, and to rest in it: in fact the former is natural for any natural body, the latter is natural for no natural body.
9. Secondly: that nature likewise is said to be principle of movement and rest connectedly [copulative], if rest is intended fundamentally and not formally; and this is so if rest is

[1] 'The heaviest is what is underneath all bodies which go downwards, the lightest is what is above all bodies which go upwards.' All translations from Greek are mine. I reproduce the Greek text as it appears in the text. The only change is the adoption of our contemporary style for the characters.

[2] 'We call light absolutely what is drawn upwards and towards to extreme, heavy absolutely what is drawn downwards and towards the centre.'

[3] 'By nature every body is drawn to and remains in its own place.'

understood as the very possession and fruition of the form and of the terminus, not simply as privation of movement.

10. And in this sense, κινεῖσθαι καὶ ἵσασθαι [to move and to stop] is the same as κινεῖσθαι καὶ ἠρεμίζεσθαι [to move and to rest], according to Aristotle, 6 Phys. 8 text. 67.

11. And also in this sense, very acutely Aristotle said that εἰς τὸν αὐτοῦ τόπον φέρεθαι ἕκασον, εἰς τὸ αὐτοῦ εἶδός ἐσι φέρεσθαι. 4. Coel. 3.[4]

12. Finally, in the same sense, nature is not only the efficient cause, but also the end of every natural movement. ἡ φύσις ἡ λεγομένη ὡ γένεσις, ὁδός ἐσιν εἰς φύσιν 2 Phys. 1. text. 14.[5]

III. Α"γήρατον ἀναλλοίωντον καὶ ἀπαθές ἐσι τὸ πρῶτον τῶν σομάτων. I *de Coel. 3. Text. 22.*[6]

1. According to Aristotle, the nature of the heavens is only the principle of natural movement in itself.

2. Only the local movement is reciprocal with nature, and with the natural body in general.

3. Then, movement generally taken is not an affection of the natural body.

4. In the definition of nature, we must intend especially local movement, not movement in general.

5. The act of the mobile as mobile is the definition of movement in general; anyways, it is not the definition of the affection, unless we limit it to local movement.

6. Movement in general is reciprocal with the natural body not differently as sensitive is with animal, because of the unique tactile [property].

IV. *Infinite is* οὗ ἀεί τι ἔξω ἐσί [*something beyond which there is always something else*]. *3. Phys. 6. text. 62.*

1. So, infinite is properly without an intrinsic boundary.

2. Consequently, because of the law of the opposites, every finite will include an intrinsic boundary at both ends.

3. By which means, every motus with respect to finite magnitude, is finite and continuous; every motus (as much as all the other continuous things) has intrinsical boundaries not only in its end, but also in its beginning.

---

[4] 'To be drawn towards its own place is like to be drawn towards its own form.'

[5] 'Nature is said to be, as generation, a path towards nature.'

[6] 'The first body is ungenerated, incorruptible and not passive.'

4. Therefore, 'being changed' is not simply the terminus of movement.

5. Neither το φερόμενον [the moved body] nor διαίρεσις κινήσεος [division of movement], truly are proper and simple termini of movement: in fact the former is only assigned to local movement, the latter to any movement indeed, but only by metonymy.

6. Therefore, as we are instructed by Aristotle himself in 6. Phys. 10. text 88., we are perhaps among the first (said without malice), to call κινηματα (in plainer and clearer way) the boundaries of movement simply indivisible, which respond to the name of point in a line and moment in time.

V. *It is well known by nature that heavier bodies are everywhere in the universe below lighter bodies, as it is well know by experience.*

1. It follows that the elements in the middle, in particular those which are [a mixture of] heavy and light, can gravitate and levitate in every place, in their place as much as in that of the other elements; when the other elements are removed, or when they are driven by a stronger force out of their natural place.

2. Not only water but also air can be drawn spontaneously both to the centre [of the universe] and to the heavens, also without fear of the void; and not only in order for them to be mixed together, but also while they exist in a pure form.

3. The elements can be driven (beyond their nature but yet by natural inclination) not only to fill the void, but also to drive out the plenum.

4. Hence, the potency incorporated in the elements is double: one special for their preservation, another universal and obedient [oboedientialis] for the preservation of the universe: through the first one they seek a definite and special place: through the second one they seek no definite place in the universe; yet, they observe this inviolable rule that, in order to preserve the order of the universe, the heavier bodies are below the lighter ones.

5. Accordingly, air is not moved upwards towards its place by one form, nor downwards towards the place of earth by another form, if there is either void or fire; on the contrary, it is moved by the same one according to a different end, one for its own preservation, the other for the preservation of the harmony, order and union of the universe.

VI. Ε"αν τις μεταθῇ τὴν γῆν οὗ νῦν ἡ σελήνην, οὐκ οἰσθήσεται τῶν μορίων ἕκαστον πρὸσ αὔτήν, ἀλλ'ὅπου περ καὶ νῦν. *4. de Coel. 3.*[7]

1. Parts do not move towards the whole, but towards the natural place.

[7] 'If we put the earth in the place of the moon, every part of the earth would not go towards the moon, but towards where it already goes now.'

2. Thus, the natural place has got a force [vim] to attract and preserve the thing which is in place.
3. Since the place of the heavy bodies is the centre of the universe (which is that dot in the middle of the universe which can be understood only through imagination, or otherwise nothing is really distinct from magnitude); this centre is said to attract towards itself and to preserve the heavy bodies.
4. Undoubtedly after a more subtle scrutiny, this descent of earth to that point must seem more remarkable than the approach of a sword to the stone of Heracles: and yet we ask (as the very subtle Scaliger says) by which cause a thing attracts another to itself, like iron to a magnet: instead, by which means earth is driven towards something which is nothing, we do not ask.
5. Therefore, in many things, not the subtlety of the thing itself in nature is the cause for admiration, but our own stupidity.

VII. *Especially among oviparous animals we know by experience that the eggs of females which are carried by males generate baby birds, not because of the incubation of males, but because of that of females, within the same species as much as among different species: like among hens, geese, ducks and so on. I. de gen. animal. 21.*
1. Hence, with regards to natural generation, one may infer various things: first, if the generation is supposed to be only in the production of baby birds, the seminal virtue (as they say) of the parents, especially of the male parent, is efficient to the extent that it regards the very production of form.
2. Secondly, that the females provide some sort of matter, for example they give aliment, and they keep the seed warm with their natural heat, in favour of the formation of the foetus.
3. Thirdly, generation is said to be univocal not with regards to the animal which broods or gives birth, but with regards to the animal which gives the seed, specifically the male.
4. The animal which incubates and extrinsically broods can be said to be an equivocal cause.
5. Not every equivocal cause is necessarily more noble than the effect by dignity and perfection.
6. Neither every equivocal cause includes the perfections of other species by virtue and eminence, but only the common and celestial equivocal cause does.

VIII. *In the generation of living creatures, seed is the material and the efficient cause as much as a craftsman is, according to Aristotle. I. Phys. 7. text. 2. 2 Phys. 3. text. 31. Metaph. ζ. θ. text 31. I de gen. anim. 21.*

1. Thus, the seed is not a uniform body, but it is made of different parts, some more subtle, some others rougher; with respect to the more subtle ones it acts as a craftsman: with respect to the rougher ones, is passive [patiens], just like matter.

2. Since the natural generation of living creatures, in plants and lower animals, is without any doubt univocal, it is, in generation, the principal cause, not an instrument.

3. Thus, the seeds of plants and lower animals are animate in act.

4. Hence, with Scaliger, very trained [exercitatissimo] in the subtleties of nature,[8] the spirit [animus] is willing to confirm many almost paradoxical points. First, the seed of oil, is oil: and the seed of a dog, is a dog; although imperfect, lacking only a jointed structure.

5. Secondly, the form of a dog can be said to be in the potency of the seed itself, since the seed is able to convey [potens] the form of dog it contains in itself.

6. Thirdly, the form of a dog is educed out of the potency of the seed, not with respect to the first act but precisely with respect to the second one. In fact the very form pre-exists, therefore the outcome is not the form itself, but its act, thanks to which it can thereafter exert itself.

7. Fourthly, the first actions of the soul in the seed, which follow closely from its potency, are the disposition and conformation of its limbs, in order to receive in conformed and well-disposed limbs later and more perfect actions, as the operations and the senses.

8. The soul of the seed, without the instrument of its location [domicilij], is architectonic [architecta]. In fact, no quality known to man can be the instrument of the ordination, location [situs], number and shape of the parts of the organic body: although qualities can be the first instruments of secretion or condensing, of condensation or rarefaction, of extension or contraction, of roughness or smoothness, of hardness or softness.

9. Purely natural generation is not the production ex novo of some form, but only the reduction of form to act, or better, its promotion to the production of effects.

10. So, the tree generates as soon as the seed produces, it is not instead generated when it sprouts from the seed. Thus, the dog is not generated when a puppy is born, but when the seed sprouts.

11. Neither it follows from this that the dog is fully subject to several souls, since it has only one, which is enough to the generation of many souls: like in the branches of oil in which there are many parts, a single soul is the one from which many come forth.

12. And as one single soul, in the increments, puts on in the aliment a new and multiple matter, it informs the same matter and it is united with the pre-existing matter; that is why the same soul, which is just material, cannot move itself forward in the generation towards many matters.

13. In whatever way all the things said above are possible, anyways we conclude that the rational soul, according to Aristotle more divine than the others, does not propagate itself in the seed this way, but the deficiency of the generating soul in the seed is compensated by the immediate action of God, in the formation of the body as much as in the creation of a new form.

14. And according to these premises, we reject by faith the truth of this proposition, *a man can generate another man*, although almost everybody cry out in protest; and a good many, who rant on about it in its defence, miserably torment themselves.

[8] Reid is playing with the title of Scaliger's *Exoticarum Exercitationum Liber XV de Subtilitate, ad Hieronimum Cardanum*.

4. *Theses physicae*, J. Dalrymple, 1646

Complete title: *Theses logicae, metaphysicae, physicae, mathematicae, et ethicae*, Glasgow 1646, printed by George Anderson. This set of theses was discussed on July 27th 1646, 'συν θεω, *publice, in communi Gymnasij Auditorio hora solita*'.

James Dalrymple, first Viscount Stair (1619-1695) was a lawyer, a philosopher and a politician. He joined Glasgow University in 1633 and graduated in 1637. We unfortunately have no graduation theses for the student years of Dalrymple in Glasgow. From his departure from university to his appointment as regent in 1641 Dalrymple spent some time in the army fighting against the king in the first Bishops' War (1639). Dalrymple's first appointment was for a fourth class teaching (Greek and dialectic), renewed in 1642 when the regenting system was revived. The 1646 graduation theses were written for the 1643 class. After leaving university in 1647, Dalrymple moved on to a legal career. We are informed by the Oxford Dictionary of National Biography that Dalrymple did not study law abroad: this information is helpful in the assessment of his philosophy as well. If he did not study abroad, we can assume that his philosophical formation was acquired in Glasgow.

His legal and political career (culminated in a number of key roles played in Scottish political life and in the composition of the *Institutions of the Law of Scotland*, 1661) Dalrymple left the country in 1682 for the Low Countries, due to political reasons and threats to his life. There he published the *Institutions* (1681) and the *Physiologia Nova Experimentalis*, Leiden 1686. Dalrymple eventually returned to Scotland, as a supporter of the 1688 revolution, and engaged in politics again.[1]

His set of theses is particularly important for three reasons: 1) it is the only one extant for the University of Glasgow in the first half of the seventeenth century; 2) the regent brings forward some innovative theories in the context of Scholastic philosophy [which I analyse in part I, chapter 2, section 2.2], and his theses are in general among the most detailed in Scotland; 3) J. Dalrymple, later Viscount Stair, member of a distinguished Glaswegian family, was later in life raised to a public role in Scotland, and made important contributions to Scottish law and philosophy.

[1] I take this information from the *DNB* entry.

Dalrymple's graduation theses are an insightful case of late Scottish Scholasticism arguably influenced by some themes of early modern philosophy.

Thesis I. Φυσις ἐσιν ἄρχη και αἴτια του κινεισθαι, και ἤρεμειν, ἐν ω ὕπαρχει πρωτως, καθ αυτο, και μη κατα συμβεβηκος,[2] Arist. ch. I. bk. 2. Nature thus defined does not include the essences of other beings (which are the principles of the changes [mutationum] and of the properties which belong to the beings in a state of rest) any less than it includes the essence of the body, unless some reason to say the contrary can be given; nature is therefore restricted to the essential parts which constitute the body, those which, in particular, deserve the name of nature; and the very body is called natural, and the whole science is called Physics or natural science: science which carefully considers the body, in itself and in its species absolutely considered, by investigating its Nature and demonstrating its affections by Universal principles.

II. There are three principles of the natural body in becoming [in fieri]: matter, form and privation; in fact there are two which take on the name of nature, matter and form, of which the latter is the active principle, the former the passive principle; it is possible to freely assign the name of nature either synonymically or analogically, or jointly and collectively, or separately and distinctively.

III. Prime matter is the subject out of which a thing becomes [fit], and which endures in every mutation; it is ungenerable and incorruptible, neither is it now something different from matter when it was first created.

IIII. Matter as such is bare of all form, thus it lacks every Physical act; yet, since a being is similar to other beings in some degrees of entity, and is different in some others, matter does have a distinctive [differentialem] entitative act, in virtue of which it is different from the other beings; which act can be very well understood as pure passive potency open to the reception of form, neither can anyone deny that pure physical potency is a Metaphysical act.

---

[2] 'Nature is the principle and the cause of movement and rest, in everything that exists first per se and not by accident.' All translations from Greek are mine. I reproduce the Greek text as it appears in the text. The only change is the adoption of our contemporary style for the characters.

V. By an innate appetite, prime matter strives for all forms without distinction: not with the particular regard to what is more perfect or less perfect, or to this or that, but with the simple regard to form; thus neither form is unwillingly retained by matter, nor matter attempts to reject a form in order to receive another one; therefore, prime matter is falsely accused of being the origin of corruption, because it concurs just in a passive way.

VI. Matter lacks any activity and efficacy, unless it is raised as instrument of something else.

VII. Physical form is a substance really different from matter, and just as matter is pure potency, form is pure Physical act, which does not include anything material; yet, which naturally requires it as partner for its own good, or which necessarily demands it in order to be preserved and operate; hence this form is called material, the other is called spiritual and immaterial.

VIII. When Aristotle defines form as λογον της οὐσιας,[3] he is not talking about the Physical form (in fact he is dealing with causes in general), but about the Metaphysical form, which is very appropriately called a 'formality', from which the specifications of all things arise; specifications which are nevertheless taken on remotely by the physical form.

IX. The natural bodies are not specified by form with respect to their particular entity, but by reason of their nature; namely, as it is the principle and root of the different affections impressed on itself by the agent, and of the operations thence emanating: so, there are not as many different forms as species of bodies, and the species can vary while the form remains the same; neither is it necessary to imagine in the perpetual course of generations that new forms arise from non-existing things.

X. To ascribe the origin and the production of forms to the eduction from matter leads to a hopeless entanglement; because the potency of matter is wholly ineffective, just passive and receptive, and the eduction often takes place by Instrumental cause, or by a cause inferior to the effect that is to be produced.

[3] '[formal] reason, essence of the substance'.

XI. We ascribe the production of forms to GOD, and we ascribe the propagation to the union or to the disjunction of the same produced form, or to the specification of the different impressed affections.

XII. GOD in the beginning impressed one intimate bodily form on the whole mass [massae] of matter, form from which matter is established as a body; which form remains the same in all bodies, has no contrary form through which it is expelled, and is coeval with matter, and of equal antiquity; it is however the cause of all the various affections, by virtue of several concurring agents; therefore, sometimes it takes on one species, sometimes another.

XIII. Besides, there are other forms, which Bodiliness admits as further degrees of essential perfection; in the creation of each species of individuals, these forms are first divinely bestowed and are carried forward up to this time by continuous offspring, therefore the Generans does not bestow on the generated only a portion of matter but also a portion of form; from which the agent, by favouring and exciting them, can rouse a new individual of the same species.

XIV. In rougher bodies, any part can become a new and complete individual of the same species, either by mere discontinuity (as for water divided in several portions), or by the fertile assistance of a different agent, which confers the dispositions required to the exercise of the faculties (as for a cut-off branch, which becomes a whole tree if put into the ground). On the other hand, in more perfect bodies there is a certain part, intended by nature, which alone can be roused to the perfection of a new individual, if it is commanded and assisted by a suitable cause (as for the seed of living creatures, of plants in the ground, of animals in the womb, labouring as in a receptacle).

XV. After the destruction of an individual, the whole form does not cease at once, but, while still adhering to matter, it takes on new species, similar but less perfect, from which worms appear out of the corpses, and various little animals appear out of the flesh of different sorts of animals.

XVI. The affections of a natural body belong to it principally in reason of its matter or of its form; Quantity, Continuity, Infinity and Whereness [Ubicatio] in place are affections of the first kind: movement and duration are of the second.

XVII. We are doubtful whether Quantity first belongs to prime matter, or to the body, neither do we determine whether quantity really differs from matter, or not; nonetheless, we positively hold that quantity (whatever quantity is) is adequate to matter and immutable, neither can it increase or decrease, unless matter is added or subtracted.

XVIII. Any quantified and continuous being is infinitely divisible, but it does not include within itself actually infinite divisibles, whether in number or size [mole]; neither is there any potency out of which they can be drawn, and although it includes indivisibles as termini, it cannot be composed of either finite indivisible termini, or infinite indivisibles.

XIX. Whereness is the extension of the body as it impenetrably occupies a certain space, which is equal to it in all dimensions, and it essentially has place as a boundary, which therefore is not the surrounding surface, present only by accident, but the very space or gap, which cannot be missing at all.

XX. A surface can be called an extrinsic place by analogical attribution of an extrinsic term to the thing named; for that reason, place is defined by Aristotle as the immobile and first surface of the containing body; without excluding the space itself, but excluding the former gap, explained by it: more correctly, by claiming this surface immobile, Aristotle shows that space is the very internal place, surface is its boundary, for surface is called immobile only as boundary of an immobile space.

XXI. By its reason, a change of surface is not a change of place or local movement [latio], but a change of surface (as it is the extrinsic boundary) of an immobile space.

XXII. Quantity and Body, or impenetrability and quantity, cannot be one without the other: therefore this implies that several bodies cannot be in the same equal place at the same time.

XXIII. And a single body, as a whole, can be in different places no more and no more truly than it can be detached from itself.

XXIV. Movement is the act of a being in potency, insofar as it is in potency, not towards another act but towards its same act; and, given this definition, movement includes every changes, as well as mutations which occur in an instant; it extends not only to Physical changes but also to any other changes, unless it is limited by the intent of the Philosopher.

XXV. There are only three species of real movement, as distinct from mutation, viz. increment, alteration and local movement. In fact, generation takes place in an instant, and corruption and decrease are not real and positive mutations.

XXVI. Rarefaction and condensation are not movements towards quantity, but alterations through which the shape [figura] of a body is changed, because of the entry of a finer body through the pores and channels of the body, from which the body appears to extend further, yet, still with the same quantity.

XXVII. Alteration neither takes place by the change of all the pre-existing qualities, nor by merely a firmer rooting in the subject; it takes place by adding a further degree to the pre-existing quality, a degree which is a partial quality, similar to the pre-existing one; and it is pointless to ask whether it is of the same or of a different species, whether heterogeneous or homogeneous, since it concurs in the same numerical and individual identity.

XXVIII. The first specific division of the body is into simple and mixed; these bodies are not different because of different Physical forms, which belong reciprocally to both, but because of abstract Metaphysical formalities; it is therefore more of a Metaphysical division than of a Physical one. There is a similar division of the simple body into Heavens and Elements.

XXIX. The world has five elements, the mixed ones are four, viz. Earth, Water, Air, Fire; people can form their own opinion about whether the Element of fire is in the hollow space of the moon,[4] while keeping proportion with the rest of the Elements; in whatever manner, Fire truly is a simple Body and one among the Elements.

XXX. Single Elements require their own proportion, which is difficult to determine with precision. In reality this proportion does not consist in quantity, but in the rarity and position of the parts, which single Elements require by nature; therefore, when gunpowder is ignited and it immediately breaks out of a compressed place, and it seems to occupy a bigger place; this does not happen because the same matter, when set alight, takes on a bigger quantity, but because, when set alight, it requires by nature a dilatation and rarity of

[4] The idea is that the natural place of the element fire is an empty moon. The moon is in the first celestial sphere, but the fire can still be in a relation with the other sublunar elements.

the parts, caused by the permeation of a finer Body, and it breaks forth to receive it, and, once it is received, it lights up by its own light; and even if it is one undivided body, there are in fact two.

XXXI. The heavy Elements move downwards with a natural impetus by an internal principle, towards the centre of the universe, and not towards their own Element; hence, the whole Earth, once placed close to the sphere of the moon, would go down towards the centre; or, while the earth is kept there, just one particle of earth falls downwards, it does not lose its appetite, as it stays still thanks to its own Element's intervention; yet, impulse and compression remain, sideways and downwards; but, since it is completely surrounded by bodies that have equal effectiveness, it does not gravitate in its place with respect to the adjacent body.

XXXII. Anyone can choose according to their own liking whether it is possible for a light element to naturally move away from the centre, and for it to naturally tend towards an end where it rests; or rather whether it is called light because it is just less heavy, and because it is pushed upwards by the pressure [compressione] of a heavier element.

XXXIII. The mixture is the union of contrary mixables [miscibilium alteratorum], not by confusion or continuation, but by one common form or formality, in which, as in a copula, all the material parts of the mixture come together.

XXXIV. The production of an animate body is a mixture, neither is it required for the concept of a mixture that mixable elements be pure or separate immediately before generation.

XXXV. In the animate body, the form of the mixture is not different from the bodiliness and the soul; on the contrary, mixables are united in it in the unity of one body and one species.

XXXVI. The soul is the first act of an organic body, which has life in potency; therefore, it is distinguished in it from other forms, because it has different vital faculties and distinct bodily instruments by which the faculties are exercised.

XXXVII. It is a stupid figment to say that there are several souls of one single animate body, either successive or concurrent; rather, one single soul includes in itself and exerts all the faculties.

XXXVIII. The faculties of the soul neither are simply the same with their substance, nor is there are a large diverse multitude of faculties, such that so many are the external senses, so many the internal, so many are the faculties of intelligence, agent and patient; rather, in a moderate way, we distinguish one faculty of knowledge, appetite and movement from the very substance of the soul, and we refer all the faculties to those just mentioned: it is open to anyone to call them faculties that are distinct to a certain extent or to call them one faculty that is exercised in different ways.

XXXIX. There is no sensation in the external sensorium which is distinct from perception in the common sense; instead, there is one apprehension which, when the sensorium is affected, apprehends the object as present, and therefore is called external, because an external organ is affected.

XL. Sensation requires a sensible object, the sensorium, and an impression made by the object on the sensorium, which represents the object, and the impression is therefore called an impressed species. The natural sympathy between the soul and the body during their union is the reason for the stimulation of the soul to drawing out the *notitia*,[5] in such a manner that whatever the body undergoes, something similar is represented in the soul, from where also the senses of joy, pain etc. originate.

XLI. The instrument of this sympathy is the brain because, when it is affected in various ways, the apprehension is similarly altered: like when the spirits in the brain are at rest because of its fullness, and sleep follows, even if the objects produce a species in the sensorium, there is no perception.

XLII. Many useless questions are asked about the reception of the species in the soul, or about the movement of the impressed species by animal spirits as vehicles, from the sensorium to the brain, or about the illumination of the *phantasmata* by the Agent Intellect, and about the production of new intelligible species, the impression in the passive Intellect, or their conservation in memory as in a repository.

---

[5] '*Notitia*' is something *notus* [known] to us, the final result of the process of knowledge.

XLIII. We claim that the Soul has only one faculty of knowledge, through which it perceives by its natural sympathy the object which affects the sensorium, as it affects it, and draws out its representation; which is an act of simple apprehension, neither do we recognise any other expressed species. Afterwards other acts about the same object follow, which acts, while they do not transcend the perfection proper to beasts, are called *Phantasmata*: then, purer ones follow, which are called Intelligible Species, Acts of Intellect, mental Terms; by which the faculty is helped at drawing out similar acts: and from all these different operations, the senses, the Phantasies, Intellect and memory, one faculty of knowledge comes out.

XLIV. The faculty of appetite, as it tends towards less noble objects, and moves the spirits with a stronger impetus, and alters the body, is called Sensitive Appetite: instead, as it is about more sublime objects in a purer way, it is called Will.

CPSIA information can be obtained
at www.ICGtesting.com
Printed in the USA
BVHW072117120223
658291BV00014B/2073

9 781805 240723